BIOACTIVE CARBOHYDRATE POLYMERS

Proceedings of the Phytochemical Society of Europe

Volume 44

*The titles published in this series are listed at the end of this volume.*

# Bioactive Carbohydrate Polymers

*Edited by*

Berit S. Paulsen

*University of Oslo,*
*School of Pharmacy,*
*Oslo, Norway*

KLUWER ACADEMIC PUBLISHERS

DORDRECHT / BOSTON / LONDON

A C.I.P. Catalogue record for this book is available from the Library of Congress.

ISBN 0-7923-6119-9

Published by Kluwer Academic Publishers,
P.O. Box 17, 3300 AA Dordrecht, The Netherlands.

Sold and distributed in North, Central and South America
by Kluwer Academic Publishers,
101 Philip Drive, Norwell, MA 02061, U.S.A.

In all other countries, sold and distributed
by Kluwer Academic Publishers,
P.O. Box 322, 3300 AH Dordrecht, The Netherlands.



*Printed on acid-free paper*



Printed in the Netherlands.

# Contents

# Contributors

S. Alban; Institute of Pharmacy, University of Regensburg, Universitätsstrasse 31, 93040 Regensburg, Germany

C. Boisson-Vidal; Laboratoire de recherches sur les Macromolecules, CNRS, UMR 7540, Institute Galilee, Universite Paris-Nord, Av. J.B.Clement, 93430 Villateneuse, France

F. Chaubet; Laboratoire de recherches sur les Macromolecules, CNRS, UMR 7540, Institute Galilee, Universite Paris-Nord, Av. J.B.Clement, 93430 Villateneuse, France

L. Chevolot; Laboratoire de Biochimie et Molecules marines, IFREMER, B.P. 1105, Rue de lÍle d`Yeu, 44311 Nantes cedex, France

P.J.H. Daas; Department of Food Technology and Nutritional Sciences, Food Science Group, Bomenweg 2, 6703 HD Wageningen, The Netherlands

R.A. Dalmo; Institute of Marine Biochemistry, Norwegian College of Fishery Science, University of Tromsø, N-9037 Tromsø, Norway

P. Durand; Laboratoire de Biochimie et Molecules marines, IFREMER, B.P. 1105, Rue de lÍle d`Yeu, 44311 Nantes cedex, France

T. Espevik; Institute of Biotechnology and Insitutte of Cancer Research and Molecular Biology, Norwegian University of Science and Technology, N-7005 Trondheim, Norway

T. Flo; Institute of Biotechnology and Insitutte of Cancer Research and Molecular Biology, Norwegian University of Science and Technology, N-7005 Trondheim, Norway

G. Franz; Institute of Pharmacy, University of Regensburg, Universitätsstrasse 31, 93040 Regensburg, Germany

Ø. Halaas; Institute of Biotechnology and Insitutte of Cancer Research and Molecular Biology, Norwegian University of Science and Technology, N-7005 Trondheim, Norway

K. Ingolfsdottir, ; Department of Pharmacy, University of Iceland, Reykjavik, Iceland

J. Jozefonevicz; Laboratoire de recherches sur les Macromolecules, CNRS, UMR 7540, Institute Galilee, Universite Paris-Nord, Av. J.B.Clement, 93430 Villateneuse, France

S. kraus; Insitutte of Pharmacy, Pharmaceutical Biology, University of München, Karlstr. 29, D-80333 München, Germanty

J. Laencina; CEBAS-CSIC & Univeristy of Murcia, Department of Food Technology, 301'00 Murcia, Spain

D. Paper; Institute of Pharmacy, University of Regensburg, Universitätsstrasse 31, 93040 Regensburg, Germany

S.G. Ring; Department of Biochemistry, Institute of Food Research, Norwich Research Park, Colney, Norwich NR4 7UA, UK

J.A. Robertson; Department of Biochemistry, Institute of Food Research, Norwich Research Park, Colney, Norwich NR4 7UA, UK

J.M.Ros; CEBAS-CSIC & Univeristy of Murcia, Department of Food Technology, 301'00 Murcia, Spain

P. Ryden; Department of Biochemistry, Institute of Food Research, Norwich Research Park, Colney, Norwich NR4 7UA, UK

A,B. Samuelsen; Department of Pharmacognosy, School of Pharmacy, University of Oslo, P.O.Box 1068 Blindern, N-0316 Oslo, Norway

H.A. Schols; Department of Food Technology and Nutritional Sciences, Food Science Group, Bomenweg 2, 6703 HD Wageningen, The Netherlands

G. Skjåk-Bræk; Institute of Biotechnology and Insitutte of Cancer Research and Molecular Biology, Norwegian University of Science and Technology, N-7005 Trondheim, Norway

A.G.J. Voragen; Department of Food Technology and Nutritional Sciences, Food Science Group, Bomenweg 2, 6703 HD Wageningen, The Netherlands

H. Wagner; Insitutte of Pharmacy, Pharmaceutical Biology, University of München, Karlstr. 29, D-80333 München, Germanty

H. Yamada; Oriental Medicine Research Center, The Kitasato Insitutte, okyo, 108-8642, Japan

# Acknowledgements

We are indebted to the following companies and institutions for their generous support during the symposium on "BIOACTIVE CARBOHYDRATE POLYMERS" held at Sundøya, Norway, 13.-16. September 1998

**COST Chemistry**, EU
**Nycomed Pharma, a.s.**, Norway
**Swets & Zeitlinger publishers**, The Netherlands
**The Norwegian Association of Proprietor Pharmacists**, Norway
**The Norwegian Research Council**, Norway
**Thieme Publishers**, Germany

# Acknowledgements

We are grateful to the following companies and institutions for their generous support [illegible] [illegible] [illegible] [illegible] [illegible] [illegible]

COST Chemistry, EU
[illegible]
[illegible]
The Norwegian Association of [illegible]
The Norwegian Research Council, Norway
[illegible]

# Preface

This volume is compiled of contributions from 11 scientists presenting either lectures or posters at the International Symposium of the Phytochemical Society of Europe entitled "Bioactive Carbohydrate Polymers" held at Sundøya, outside Oslo, Norway, from 13. to 16. September 1998. Present at the meeting were more than 50 participants from 19 different countries and the meeting provided a good opportunity for those interested in the fieldd of bioactive carbohydrates to meet and discuss future aspects of this growing field of science. The lectures dealt with bioactive polysaccharides from plants traditionally used in Japan and China, USA, Iceland and in other European countries. Substances from micro and macro algae were described, and the uses of the polymers ranged from woundhealing in human to immunestimulation in fish. The lectures covered most aspects of importance for studying biologically active polysaccharides, both chemical, enzymatic, chromatographic as well as various biological test-systems.

I am most grateful to all the contributors to this volume for their efforts both during the meeting and also afterwards in connection with finishing the proceedings from the meeting, and I will also like to thank PSE for wanting this meeting to be arranged in Norway.

Oslo May 1999-09-30
Berit Smestad Paulsen

Chapter 1

# News on immunologically active plant polysaccharides

H. WAGNER and S. KRAUS
*Institute of Pharmacy, Pharmaceutical Biology, University of Munich, Karlstr. 29, D-80333 Muenchen, Germany*

Key words: polysaccharides, labeling, ELISA, mechanisms, antiinflammatory

Abstract: The medicinal use of plant polysaccharides as immunostimulatory, antitumoral or antiinflammatory agents has been hampered by the lack of bioavailability studies, by the scarce information available on the mechanism of action as well as the lack of studies on structure activity relationships in this class of compounds. Investigations of the possible mechanisms of action behind the noted inflammatory activity of polysaccharides have revealed that, beside influence on the complement cascade, endocrinal functions and cytokine induction, the effect on chemotaxis of leucocytes has to be considered as one important factor. In order to clarify the mechanism in pharmacological and clinical trials binding and resorption studies are needed. As a prerequisite for such studies we have developed methods for radio-, fluorescent- and gold labelling of polysaccharides. We also present the most recent results on production of polyclonal or monoclonal antibodies against polysaccharides, which can be used for the detection and quantification of polysaccharides in biological material and fluids.

## 1. Introduction

The rational use of polysaccharides isolated from plants as immunostimulatory, antitumoral or antiinflammatory agents was not possible previously, mainly due to:

- problems combined with the isolation and purification of sufficient amounts of pure bioactive polysaccharides in a reproducible manner;



*B.S. Paulsen (ed.), Bioactive Carbohydrate Polymers,* 1–14.

- unknown structure-activity relationships, as well as lack of information concerning the exact mechanism of action for the noted immunomodulatory/pharmacological activities;
- and lack of information concerning pharmacokinetics and bioavailability after *p.o.* and *parenteral* administration.

Due mainly to the first two problems listed above, major clinical studies with plant polysaccharides have hardly been possible. One exception has been Lentinan, a fungal β-1,3-glucan having antitumoral activity, which has been investigated in some clinical trials. In order to solve problems connected with the isolation and purification of plant polysaccharides, their production by cell cultures has been initiated. As has been proven by our laboratory for *Echinacea* polysaccharides, the use of the biotechnological route is possible [1] and also manageable on a larger scale [2], but has been associated with several handicaps. This is due mainly to the fact that the polysaccharides isolated using this technology differ from those obtained directly from the plant. Plant polysaccharides differed in molecular weight, sugar composition, mode of sugar interlinkages and immunological activities from those isolated using cell cultures. Additionally, large-scale isolation from cell cultures requires a large-scale fermentation equipment with a capacity of 2000 l or more. Hence, the routine biotechnological production of polysaccharides from plant cell cultures on an industrial scale has so far not yet been achieved.

The difficulties concerning studies on mechanisms of action and the bioavailability of polysaccharides after parenteral or oral administration arise from the fact that various mechanisms are possible. Possible targets can be macrophages, T-lymphocytes, NK-cells, the complement cascades or a combination of two or more of these. One possibility to overcome these problems and enable the visualization of interactions between polysaccharides and the target molecules could be the use of labeled polysaccharides and/or specific ELISAs.

## 2. Labeling of polysaccharides by radio-, fluorescence and gold markers

The major problem concerning the synthesis of labeled bioactive polysaccharides for use in various immunological techniques is to find a suitable method, *i.e.*, a method that does not interfere with the biological activity and allows a highly sensitive detection of the polysaccharide after application to a living organism. With the exceptions of the non-radioactive iodine labeling and a fluorescence labeling procedure for glucans there are very few standard methods available which are applicable on all types of polysaccharides. In order to tackle this problem we compared methods used for the labeling of other polymers.

Fig. 1a

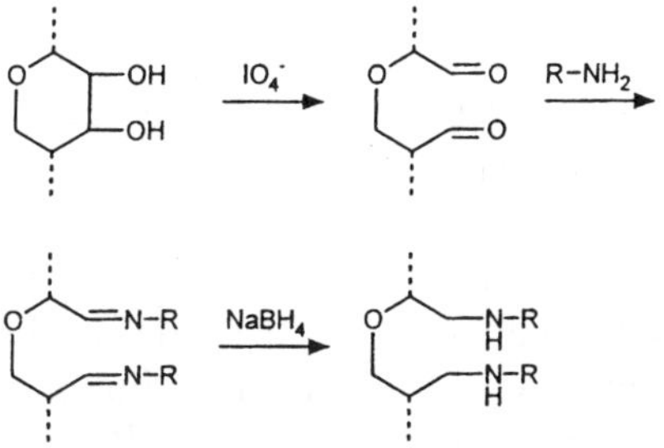


1b

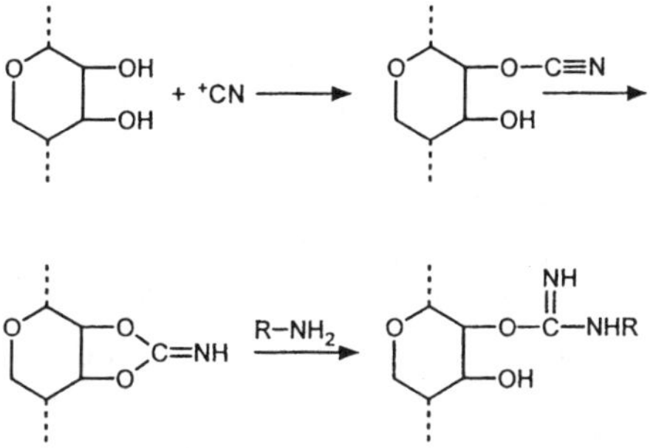


1c

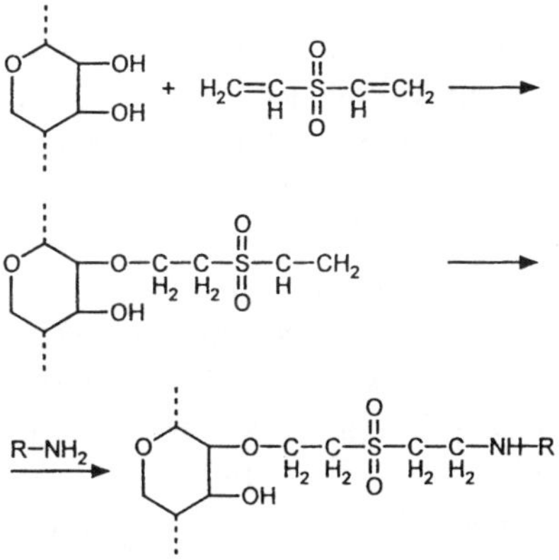


1d

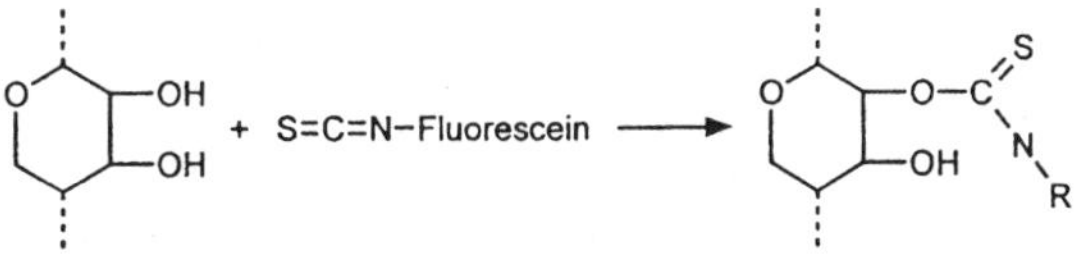


Fig. 1 Different reactions for labeling polysaccharides.

- In the first method hydroxyl groups are oxidized with periodate or dimethylsulfoxide, and the thus generated carbonyl groups transformed into Schiff's bases using an amine-marker, then into secondary amines through reductive amination [3]. This method allows to incorporate fluorescent dyes containing amines or Iodine$^{131}$-labeled tyramine (Fig. 1, a).
- In a second method, the hydroxyl groups are transferred into bromocyane or divinylsulfone derivatives, respectively, followed by the introduction of amine markers [4,5] (Fig. 1, b,c).
- In a third method, used for labeling proteins and antibodies, fluorescein isothiocyanate is introduced [6] (Fig. 1,d).

The usefulness of a method depends on its level of incorporated markers and to what extent the degradation of the polysaccharide occurs: High level of incorporation facilitates detection, whereas a low degree of alteration of the molecule or degradation ensures that enough free hydroxyls are left for binding reactions. Since Pfitzner and Moffat [7] showed that DMSO can also oxidize monosaccharides, we replaced periodate with DMSO. Using DMSO/acetic anhydride, 5 polysaccharides (2 dextrans, laminarin, larch-arabinogalactan and a neutral *Urtica*-glucan) were oxidized and the resulting carbonyl functions converted into the corresponding Schiff's bases with fluoresceinamin, Rhodamin B-amine and tyramine. Incubation with sodium borohydride produced the desired stable secondary amines. Control experiments with Concanavalin-A-sepharose (for the *Urtica*-glucan and dextran) and *Ricinus communis* agglutinine-agarose (for arabinogalactan) were used in order to investigate whether the obtained fluorescence labeled polysaccharides were still suitable for binding studies, *i.e.*, still contained intact glucan or arabinogalactan units. As confirmed by fluorescence microscopy all fluorescence-labeled polysaccharides bound to Con-A-sepharose were easily detectable. For measurement of tyramine incorporation, UV-absorption at 278 nm was used. After 120 min of incubation, a maximal degree of substitution (1 tyramine molecule per 21 sugar units) was obtained, indicating 5 - 6 tyramine units/arabinogalactan (mol.wt. ca. 20,000 D). Labeling using fluoresceinamine and Rhodamin B-amine was 5 - 6 times less effective, probably due to steric hindrance. When comparing the most optimal methods of oxidation (DMSO and periodate) using HPGPC-chromatography and RI or UV-detection respectively, we found that the elution profile of the product resulting from tyramine incorporation after oxidation with DMSO was not altered in comparison with native arabinogalactan. Labeled arabinogalactan resulting from oxidation with periodate, however, showed a drastic change in molecular weight distribution, indicating a high level of degradation. In order to investigate in what position the oxidation with DMSO had taken place, and where the molecule had been labelled, the $^{13}C$-NMR technique was used. These experiments showed that tyramine incorporation occurred primarily at the terminal C-6 and not at the reducing C-1 end. The third method, FITC-labeling was performed using dry DMSO according to the method of De Belder and Granath [8]. The highest degree of FITC-substitution (1/31) was achieved after 18 hrs of incubation. In the experiments performed by Winchester *et al.* [9] as well as those of Thornton *et al.* [6], the degree of substitution at aqueous condition reported was only 1:>1000. In order to investigate if the bioactivity of a polysaccharide was retained after FITC-labeling granulocyte chemotaxis, experiments were performed using native chemotactic *Urtica*-polysaccharide, UPS I (α,1,4-D-glucan), as well as labeled UPS I [10]. FITC-labeling reduced the chemotactic effect to ca. 55% as compared with the native polysaccharide. Using flow cytometry, we investigated if the FITC-labeled glucan binds specifically to an isolated granulocyte-fraction. Even after addition of a 100-fold excess of non-labeled glucan, it was not possible to displace the fluorescence signal of labeled glucan. Since the binding is not specific, the use of this labeling method will not enable UPS I binding studies. Another method for labeling biomolecules is the use of direct or indirect gold labeling [11,12]. Due to the high electron density of gold, gold conjugates are very suitable for detection with electron microscopy and, after silver enhancement, with light microscopy. For gold-labeling of the polysaccharide we chose an indirect two step method followed by silver enhancement. FITC-dextran was covalently bound to epoxy-active agarose and the FITC-dextran-agarose particle was incubated with anti-FITC-immunoglobulins (Fig. 2). After addition of silver reagent, the labeling was evaluated using an Epipolarisation microscope. The advantage of the use of the latter method, is that it enables the investigation of morphological changes. In judging the various methods for their usefulness for pharmacokinetic studies, we came to the conclusion that most of the methods investigated are only of limited value for the planned studies. In comparison to the fluorescence labeling the gold labeling technique seems to be much more sensitive. One important advantage of the use of fluorescence microscopy for detection is the possibility of evaluating morphological changes.

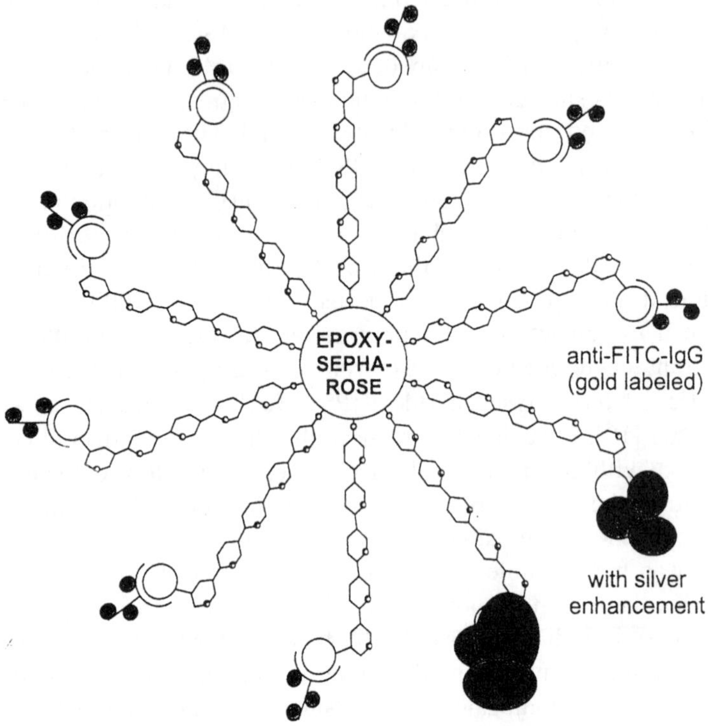


Fig. 2 Immunogoldlabeling of FITC-Dextran

## 3. ELISA-methods for labeling polysaccharides

After aller these circumstances ELISA methods might be more appropriate for pharmacokinetic and bioavailability studies. Several polyglonal ELISAs suitable for fungal polysaccharides and higher plants have been developed [13-21](Tables 1 and 2). Because, as for example with Krestin from *Coriolus versicolor*, a strong immunogenic protein part is often bound covalently to the polysaccharide, the question still remains whether the polyclonal antibodies generated are directed against sugar or protein structures. For this reason it is still unknown whether Krestin from *Coriolus versicolor,* after oral administration, is resorbed in intact form or if hydrolysis has taken place and the linked protein residue is detected instead of the carbonhydrate. We have developed polyclonal antibodies against two polysaccharides, an acidic arabinorhamnogalactan from *Echinacea purpurea* cell cultures [22] and a neutral *Urtica dioica* glucan [10]. The *Echinacea* arabinogalactan was injected subcutaneously, together with Freund`s adjuvant, into rabbits. Boosting was achieved through intramuscular injections every four weeks. For isolation and purification of the IgG-fraction, affinity chromatography was used. Through Western blot, we could assess that isolated specific IgG-antibodies were obtained. By the use of an indirect, noncompetitive ELISA, using goat-anti-rabbit-IgG antibodies, a selective detection, even in mixtures, was possible. Cross reactions were only seen with a structurally related arabinogalactan from the *Echinacea purpurea* plant. This method, detecting a minimum concentration of 1 µg/ml, was, however, not sensitive enough to enable quantitative measurements of polysaccharides in the blood after parenteral administration of 100 µg/ml. In order to characterize the antigen determinant of the arabinogalactan, we investigated the binding capacity of the terminal β-galactose-specific *Ricinus* lectin RCA 120 to the *Echinacea* arabino-galactan. We could show that RCA 120 lectin binds specifically to this polysaccharide, but it was not possible to correlate the binding capacity of the lectin to the arabinogalactan with its immunological activity. A second ELISA experiment was performed with the *Urtica* glucan UPS I. The specific IgG-antibody fraction was purified by affinity chromatography on a Protein-A column and an indirect, noncompetitive antibody capture enzyme immunoassay was then performed. The antibody titer of our α1-4-glucan was low, with a minimum concentration detected at 5 µg/ml. Cross-reaction occured with β1,3-glucans, lentinan und laminarin (40% and 20% respectively), with another glucan from *Urtica* (95%), and to the lowest extent with α1,4-glucan amylum solubile (15%). A higher selectivity can probably be obtained using monoclonal antibodies. Recently monoclonal antibodies (Mc Abs) active against a rhamnogalacturonan (RG-1) from suspension-cultured Sycamore Maple (*Acer pseudoplantanus*) cells were generated [23]. Four Mc Abs groups, which recognize epitopes on the xyloglucan part and a terminal

α-(1->2) linked fucosyl residue, were detected. Another monoclonal antibody that recognizes an arabinosylated (1->6)-β-D-galactan epitope in several plant complex carbohydrates and membrane glycoproteins has also been described by [24] (Fig. 3).

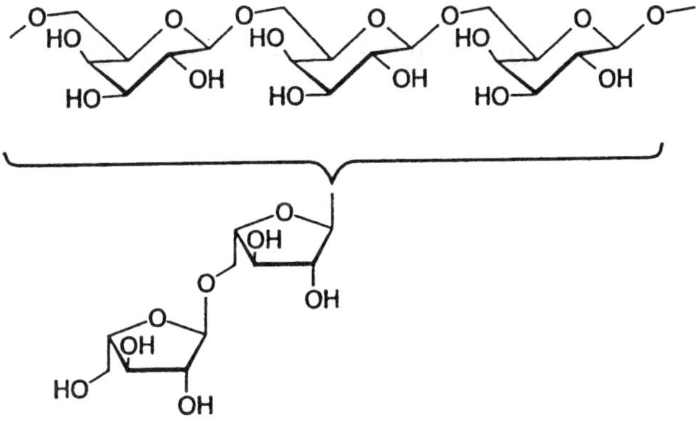


Fig. 3 Arabinosylated (1->6)-β-D-galactan epitope of some polysaccharides

Table 1

**Generation of polyclonal antibodies against polysaccharides (glycoproteins)**

| Polysaccharides (Glycoproteins) | Source | Literature |
|---|---|---|
| Schizophyllan | *Schizophyllum commun* | [13] |
| Mannan | *Saccharomyces cerevisiae* | [14] |
| Extracellular poly-saccharides | *Aspergillus* and *Penicillium* species | [15] |
| Extracellular poly-saccharides | *Ascomycets* and *Basidiomycets* | [16] |
| Extracellular poly-saccharides | *Mucor racemosus* | [17] |
| Arabinogalactan protein | *Baptisia tinctora* | [18] |
| Arabinogalactan protein | *Echinacea purpurea* | [18] |
| PSK (Krestin®) | *Coriolus versicolor* | [19,20] |
| Arabinogalactan protein | *Gladiolus* species | [21] |

Table 2

**Generation of monoclonal antibodies against polysaccharide structure**

| Polysaccharide | Source | Literature |
|---|---|---|
| Cell wall polysaccharide | *Acer pseudoplatanus* | [23] |
| Arabinosylated (1->6)-β-D-galactans | Natural and synthetic oligosaccharides, plant saccharides | [24] |

## 4. Possible Mechanism of the antiinflammatory effect of polysaccharides

The observation that several polysaccharides display significant *in vivo* anti-inflammatory activities after parenteral administration, as measured in the rat-paw edema and also partially in the rat-ear model, raises the question of the possible mechanism of action. The great structural diversity of anti-inflammatory polysaccharides, ranging from linear glucans to highly-branched acidic polygalacturonans, favors multiple mechanisms of action. From a pathophysiological point of view, many mediators liberated from immunocompetent cells and plasma enzymatic systems can be involved in the inflammatory process. The targets which can be considered for any drug interaction with polysaccharides, range from the prostaglandin metabolism, NO-mediators, and endocrinal systems (counter-irritant effect) to complement receptors, adhesion molecules, mediators of neutrophils, and leukocyte chemotaxis.

<u>Prostaglandin pathway</u>

With the exception of one polysulphated pentosan [25] an interaction of nonsulphated polysaccharides with the arachidonic acid cascade and a thus resulting decrease of pro-inflammatory prostanoids has not yet been reported. Various polysaccharides (including carrageenan and an *Echinacea* polysaccharide) were investiated for their effect on the release of arachidonic acid metabolites from stimulated and nonstimulated human PMNLs [26], however, none of the tested polysaccharides was able to activate or inhibit this pathway. In their search for the mechanism of action of the anti-inflammatory polysaccharide T-2-HN (a partially O-acetylated α-D-mannan from *Dictophora indusiata*) the authors noted only a slight influence on arachidonic acid metabolism [27]. An inhibition of the generation of vasoactive kinins (prostaglandins ?), which is postulated e.g. by Damas *et al.* [28] has not yet been reported.

<u>NO-mediators</u>

The function of nitric oxide in the carrageenan inflammation model has been investigated by several research groups. Salvemini *et al.* [29] propose nitric oxide as a key mediator in the early and late phases of carrageenan-induced rat-paw inflammation. Meller *et al.* [30] reported the participation of nitric oxide in the development and maintenance of hyperalgesia produced by intraplantar injection of carrageenan in the rat.

Counter irritant effects

Since Rocha e Silva *et al.* [31] reported that a number of sulphated polymers show antiinflammatory potential, counter-irritation has been discussed as one possible mechanism. Beside turpentine [32] and sulphated polysaccharides, neutral glucans have also been found to possess irritating pro-inflammatory activity [33]. Most likely this class of polysaccharides also exhibit their anti-inflammatory effects in the carrageenan-rat paw edema through a counter irritation effect. This is consistent with the finding of Ferreira *et al.* [34] that counter irritant effects are already able to reduce edema formation 2hs after administration of the irritating substance.

Since inhibitors of the adrenocorticoid synthesis [35] neutralize counter irritation effects, the participation of the symphato-adrenal system has been suggested. The first evidence for the involvement of neutral glucans in the induction of adrenocorticoid release was reported by Wang and Zhu [36]. A β-glucan from *Omphalia lapidescens*, which has shown anti-inflammatory activity in animal models, was able to increase the plasma content of corticosterone in rats. Beside this β-glucan, other glucans like UPS I from *Urtica dioica*, can be expected to exhibit an anti-inflammatory effect through corticoid release. The time course of corticoid release, reaching a maximum 7hs after injection of carrageenan [37] could explain why maximum of anti-inflammatory activity of UPS I was found after 24hs.

Adhesion and complement molecules

Aside from chemotactic factors generated by the activated complement cascade, complement receptors play a key role in the removal of immune complexes, lysis of bacteria, and engulfing of opsonized particles by macrophages. Most of the structure-activity relationship studies made with polysaccharides influencing the complement cascades have been performed in the laboratory of Professor Yamada [38]; however, only a few investigations have been carried out to elucidate to what extent, if at all, the complement cascades are involved in the *in vivo* anti-inflammatory activity of polysaccharides. More than 50 polysaccharides from higher plants were found to influence the classical or alternative complement pathway in the conc. range of 500 to 1000 μg/ml. In table 3 we have listed polysaccharides found in the literature [39-46], which show anti-complementary as well as rat-paw edema inhibiting activities. These polysaccharides belong to various classes of compounds but mainly glycuronans and glucans are dominating. The concentrations needed for inhibition of Carrageenan rat-paw edema by about 30 - 50% after *i.p.* or *i.v.* application range between 0.1 and 10 mg/kg. Because all polysaccharides with immunomodulating activity have not yet

Table 3

**Rat paw edema and complement inhibiting polysaccharides**

| Plant source | Type | MW | Literature |
|---|---|---|---|
| *Achyrocline satureioides* | glykanogalacturonan | ~7600/~15000 | [39] |
| *Arnica montana* | acidic heteroglycan fraction | | [40] |
| *Pinellia ternata* | glucan<br>galactan | ~15000/<br>~118000 | [41] |
| *Rumex acetosa* | rhamno-galacturonan | ~40000 | [42] |
| *Sabal serrulata* | arabino-rhamno galacturonan | ~100000 | [43,44] |
| *Sedum telephium* | polygalacturonan | ~13000 | [45] |
| *Urtica dioica* | α1,4-glucan | ~10000 | [46] |

been investigated in the rat-paw edema model, a correlation between the *in vitro* anti-complementary and edema-formation effects cannot be drawn. In this context, it might be of interest that a reduced complement level caused by the polysulphonated urea derivative Suramin did not result in an inhibition of carrageenan-induced rat-paw edema [47,48]. Therefore other immunological or non-immunological mechanisms of action, such as the role of adhesion molecules or chemotaxis must also be considered. The complement receptors play a key role in the removal of immune complexes and engulfing of opsonized particles. Complement receptor 3 (CR3 = Mac1, Integrin CD11b/CD18) possesses adhesion-molecule functions and mediates the 'sticking' of leukocytes to activated endothel, an important step in the extravazation of inflammatory cells. Because Thornton *et al.* [6] demonstrated that β-glucans bind specifically to CR3 and thus block the binding of antibodies directed against CR3, β-glucans should be able to decrease the extravazation of leukocytes by blocking cell interactions between endothel and leukocytes. S-4001, a β-glucan which matches exactly the structural requirements for binding to CR3, has been isolated from *Omphalia lapidescens*. This polysaccharide was found to have significant anti-inflammatory activity in various experimental animal models including croton oil-induced ear edema [36]. The fact that an inhibitory effect on leukocyte migration was observed indicates that the blocking of cell adhesion molecules could be another mechanism of action for the anti-inflammatory activity of this β-glucan.

Further evidence for the importance of adhesion molecules as targets for anti-inflammatory polysaccharides was provided by Dong and Murphy [49,50]. They showed, using an *in vivo* mouse model, that a glucuronoxylomannan from *Cryptococcus neoformans* inhibits leukocyte extravazation into sites of acute inflammation. They propose that binding to CD18 (part of the cell adhesion molecule Mac1) on human neutrophils is responsible for the observed inhibition of leukocyte migration.

<u>Chemotaxis</u>

It has been reported that an activation of the complement cascade results in generation of the chemotactic cleavage product C5a. Through this activation, polysaccharides were found to increase leukocyte migration in an *in vitro* chemotaxis assay when incubated in the presence of serum [51,52].

Because the glucans of *Urtica dioica* showed complement and rat-paw edema inhibiting effects, we subjected UPS I to a chemotaxis experiment using the modified micropore filter method reported previously [53,54]. In contrast to the expected inhibition of the PMNL migration, UPS I increased the migration. Other glucans (Laminarin, Lentinan) also showed stimulation of chemotaxis, but to a much lesser extent.

In contrast to this increased chemotaxis mediated by the complement cascade we found that UPS I (an α-glucan from the rhizome of *Urtica dioica* L.) is able to stimulate dose dependent leukocyte migration by itself (Fig. 4). At a concentration of 12.5 mg/ml, the chemotactic response was eight times higher as compared to a negative control [10]. Since Silva *et al.* [55] could demonstrate that, even 4 hs after an intraperitoneal administration of a β-glucan from *Paracoccidioides brasiliensis*, the polymorphonuclear cell recruitment into the peritoneal cavity of rats was strongly increased, it is possible that, as a result of this influx of leukocytes, the amount of inflammatory cells in the blood vessels is reduced and an anti-inflammatory effect is generated. This would be in agreement with the findings of Damas *et al.* [32], who showed that carrageenan edema in rats can be inhibited by leukopenia. On this basis, the anti-inflammatory effect of UPS I, a 60% reduction of the rat paw edema volume at an *i.p.* administration of 10 mg/kg, can be partially explained.

The chemotaxis experiment clearly shows that the anti-inflammatory activity of polysaccharides as measured in the rat edema model cannot be explained by one single mechanism of action. Furthermore, it is evident that the degree of an anti-inflammatory effect is strongly dependent on the secondary and tertiary structure of a polysaccharide, whether a polysaccharide has a single or a triple helix structure and a linear backbone only, or a branched structure.

## References


[1] Wagner, H., Stuppner, H., Puhlmann, I., Jurcic, K., Zenk, M.A. (1988). *Phytochemistry 27*, 11-126.

[2] Rittershaus, E., Ulrich, I., Weiss, A., Westphal, V. (1989) *Bio Engineering 5*, 28-34 *ibid*. 5, 51-65.

[3] Lachmann, U., Schmidt, W., Wunderlich, H., and Schäfer, H. (1992). *Zbl. Bakt. 276*(3): 313-322.

[4] Glabe, C.G., Harty, P.K. and Rose, S.D (1983). *Anal. Biochem. 130*: 287-294.

[5] Houen, G. and Jensen, O.M. (1995). *J. Immunol. Meth. 181*: 187-200.

[6] Thornton, B.P., Vetvicka, V., Pitman, M., Goldman, R.C. and Ross, G.D. (1996) *J. Immunol. 156*(3): 1235-1246.

[7] Pfitzner, K.E. and Moffat, J.G. (1963) *J. Amer. Chem. Soc. 85*: 3027-3028.

[8] De Belder, A.N. and Granath, K. (1973), *Carbohydr. Res. 30*: 375-378.

[9] Winchester, R.J. Ross, G.D. (1986) In: Manual of Clinical Laboratory Immunology, N.R. Rose, H. Friedman and I.L. Fahey, Eds., Am.Chem.Soc.Microbiol., Washington DC: 212-225.

[10] Wagner, H., Kraus, St., Jurcic, K. (1998). In: Immunomodulatory Agents from Plants, Ed. H. Wagner, Birkhäuser Publ. Comp., Basel, Switzerland.

[11] Segond-von Banchet, G., Heppelmann, B. (1995). *J. Histochem. Cytochem. 43*(8): 821-827.

[12] Aurion (1997), *Workshop Syllabus*, Heidelberg, DKFZ.

[13] Tabata, K., Itoh, W.; Hirata, A., Sugawara, I., and Mori, S. (1990):. *Agric. Biol. Chem. 54*(8): 1953-1960.

[14] Mikami, T., Suzuki, S., Schuerch, C., Suzuki, M. (1986): ELISA for several mannans. *Chem.Pharm.Bull* (Tokyo) *34*(9): 3933-3935.

[15] Kamphuis, H.-J. De-Ruiter, G.A., Veeneman, G.H., Van-Boom, J.H., Rombouts, F.M., Notermans, S.H.W. (1992). *Antonie Van Leeuwenhoek 61*(4): 323-332.

[16] Middelhoven, W.J., and Notermans, S. (1988). *J.Gen.Appl.Microbiol. 34*(1): 15-26.

[17] De Ruiter, G.A., Van-der-Lugt, A.W., Voragen, A.G.J., Rombouts, F.M., Notermans, S.H.W. (1991). *Carbohydr. Res. 215*(1): 47-58.

[18] Egert, D., Bodinet, C. and Beuscher, N. (1989). 37th Annual Congress on Medicinal Plant Research, Braunschweig, Germany, Sept. 5-9, 1989. *Planta Med. 55*(7): 637.

[19] Ikuzawa, M., Matsunaga, K., Nishiyama, S., Nakajima, S., Kobayashi, Y., Endoh, T., Kobayashi, A., Ohara, M., Ohmura, Y., Wada, T. and Yoshikumu, C. (1988). *Int. J. Immunopharmacol. 10*(4): 415-423.

[20] Endoh, H., Matsunaga, K., Yoshikumi, C., Kawai, Y., Suzuki, T. and Nomoto, K. (1988). *Int. J. Immunopharmacol. 10*(2): 103-109.

[21] Gleeson, P.A., and Clarke, A.E. (1980). *Biochem. J. 191*(2): 437-448.

[22] Schöllhorn, C. (1994): Strukturchemische und immunologische Untersuchungen von Polysacchariden aus *Withania somnifera* (L.) Dunal und Untersuchungen zur Struktur-Wirkungs-Beziehung von Polysacchariden aus *Echinacea purpurea*-Zellsuspensionskulturen. PhD Thesis, Faculty of Chemistry and Pharmacy. University of Munich.

[23] Puhlmann, J., Bucheli, E., Swain, M.J., Dunning, N., Albersheim, P., Darvill, A.G., and Hahn, M.G. (1994) *Plant Physiol. 104*: 699-710.

[24] Steffan, W., Kovác, Albersheim, P., Darvill, A.G. and Hahn, M.G. (1995). *Carb. Res. 275*: 295-307.

[25] Freyburger, G., Larrue, F., Manciet, G., Lorient-Roudaut, M.F., Larrue, J. and Boisseau, M.R. (1987). *Thromb. Haemost.* *57*(3): 322-325.

[26] Panossian, A.G., Gabrielian, E., Manvelian, V., Jurcic., K., and Wagner, H. (1996). *Phytomedicine* *3*(1): 19-28.

[27] Ukai, S., Hara, C. and Kiho, T. (1982). *Chem. Pharm. Bull.* *30*(6): 2147-2154.

[28] Damas, J. and Remacle-Volon (1982). *Pharmacol.* *13*(2): 225-239.

[29] Salvemini, D., Wang, Z.-Q., Wyatt, P.S., Bourdon, D.M., Marino, M.H., Manning, P.T. and Currie, M.G. (1996). *Br. J. Pharm.* *118*: 829-838.

[30] Meller, S.T., Cummings, C.P., Traub, R.J. and Gebhart, G.F. (1994). *Neuroscience* *60*(2): 367-374.

[31] Rocha e Silva, M., Cavalcanti, R.Q. and Reis, M.L. (1996). *Biochem. Pharmacol.* *18*(6): 1285-1295

[32] Damas, J., Remacle-Volon, G. and Deflande, E (1986): *Nauyn-Schmiedeberg's Arch. Pharmacol.* 332(2): 196-200.

[33] Abe, S., Takahashi, K., Tsubouchi, J., Aida, K., Yamazaki, M., and Mizuno, D. (1984):. *Gann.* *75*(5): 459-465.

[34] Ferreira, S.H., Lorenzetti, B.B. and Correa, F.M.A. (1978). *Eur. J. Pharm.* 53: 39-48.

[35] Bhattacharya, S.K., Das, N. and Rao, P.J. (1987). *J. Pharm. Pharmacol.* *39*(10): 854-856.

[36] Wang, W.J. and Zhu, X.Y. (1989). *Acta.Pharm. Sin.* *24*(2): 151-154.

[37] Stenberg, V.I., Bouley, M.G., Katz, B.M., Lee, K.J. and Parmar, S.S. (1990). *Agents and Actions* *29*(3-4): 189-195.

[38] Yamada, H. In: Immunomodulatory Agents from Plants, Ed. H. Wagner. Birkhäuser Publ. Company, Basel, Switzerland.

[39] Puhlmann, J., Knaus, U., Tubaro, L., Schäfer, W., Wagner, H.: (1992). *Phytochemistry* *31*: 2617-2621.

[40] Puhlmann, J., Zenk, M.H., Wagner, H. (1991). *Phytochemistry* *30*: 1141-1145.

[41] Zhang, D.Y., Mori, M., Hall, H., Lee, K.H. (1991). *Int. J. Pharmacognosy* *29*: 29-32.

[42] Schwartner, C. (1996): Entzündungshemmende und antioxidative Wirkstoffe aus *Rumex acetosa* L. und anderen Arzneipflanzen. PhD Thesis, Faculty of Chemistry and Pharmacy, University of Munich.

[43] Wagner, H., Flachsbarth, H. (1981).. *Planta Med.*, *41*: 244-251.

[44] Wagner, H., Flachsbarth, H., Vogel, G. (1981). *Planta Med.*, *41*: 252.

[45] Sendl, A., Mulinacci, N., Vincieri, F.F., Wagner, H. (1993). *Phytochemistry* *34*, (5): 1357-1362.

[46] Wagner, H., Willer, F., Samtleben, R., Boos,G.(1994).*Phytomedicine* *1*: 213-224.

[47] Vinegar, R., Truax, J.F. and Selph, J.L. (1976). *Fed. Proc.* *35*(13): 2447-2456.

[48] Calhoun, W., Chang, J. and Carlson, R.P. (1987). *Agents Actions* *21*( 3 - 4): 306-309.

[49] Dong, Z.M. and Murphy, J.W. (1995). *Infect. Immun.* *63*(3): 770-778.

[50] Dong, Z.M. and Murphy, J.W. (1997). *Infect. Immun.* *65*(2):557-563.

[51] Torisu, M., Hayashi, Y., Ishimitu, T., Fujimura, T., Iwasaki, K., Katano, M., Yamamoto, H., Kimura, Y., Takesue, M., Kondo, M. and Nomoto, K. (1990). *Cancer Immunol. Immunother.* *31*: 261-268.

[52] Pereira Crott, L.S., Lucisano, Y.M., Siila, C.L. and Barbossa, J.E. (1993). *Journal of Medical and Veterinary Mycology* *31*: 17-27.

[53] Wilkinson, P.C. (1988).. *Methods Enzymol.* *162*: 38-50.

[54] Zigmond, S.H. and Hirsch, J.G. (1973 *J. Exp. Med.* *137*(2): 387-410.

[55] Silva, C.L., Alves, L.M. and Figueiredo, F. (1994). *Microbiology* *140*: 1189-1194.

Chapter 2

# Bioactive plant polysaccharides from Japanese and Chinese traditional herbal medicines

H. YAMADA
*Oriental Medicine Research Center, The Kitasato Institute, Tokyo 108-8642, Japan*
**Key words**: pectins, anti-ulcer, immunomodulation, IL-6

**Abstract**: Varoius pharmacological activities have been observed in pectic polysaccharides and pectins from hot water exreact of medicinal herbs including Kampo (Japanese herbal) medicines. Therefore studies on bioactive polysaccharides are important for elucidation of efficacy of herbal medicines and development of new carbohydrate medicines. The present paper deals with recent studies on bioactive plant polysaccharides from Japanese and tradtional herbal medicines, and include results on anti-ulcer and mitogenic activity of pectins from *Bupleurum falcatum*, intestinal immune system modulating polysaccharides from the rhizomes of *Atractylodes lancea* and IL-6 production enhancing activity of rhamnogalacturonan II dimers from leaves of *Panax ginseng*.

## 1. Introduction

As Japanese herbal medicines, so called Kampo medicines originated in China more than 2,000 years ago, extensive knowledge and cumulative experience regarding their use have been acquired. In Japan, Kampo medicines are now important alternative medicine for the treatment of diseases which are not cure by Western medicines, and more than 80% of practising physicians have used Kampo medicines at one time or another [1]. Therefore scientific evaluation on the efficacy of Kampo medicines is highly required. Because Kampo medicines have been developed by clinical effects from ancient years, their basic researches have a possibility to find new drugs. Kampo medicines are generally used as the prescription, and contain substances with both low molecular weights such as alkaloids, terpenoids, saponins and flavonoids and high molecular weights such as proteins, tannins and polysaccharides in the hot water extract. Although biologically



*B.S. Paulsen (ed.), Bioactive Carbohydrate Polymers,* 15–24.

active substances with low molecular weight in Kampo medicines have been studied well, they can not account for all of the clinical effects achieved. Of the fraction with a high molecular weight, polysaccharides have been shown to possess a various pharmacological activities on immune, coagulative, digestive, circulatory and endocrine system etc [2, 3]. Therefore studies on bioactive polysaccharides are important for elucidation of efficacy of herbal medicines including Kampo medicines and development of new carbohydrate medicines. The present paper deals with recent our studies on bioactive plant polysaccharides from Japanese and Chinese traditional herbal medicines.

## 2. Anti-ulcer and mitogenic pectin from roots of *Bupleurum falcatum*

The roots of *Bupleurum falcatum* have been used clinically in Kampo medicines for the treatment of chronic hepatitis, nephrotic syndrome and auto-immune diseases. Bioactive pectin, bupleuran 2IIc, which was isolated from roots of *B. falcatum*, [4] showed potent anti-ulcer activity against HCl-ethanol-induced gastric mucosal lesions in mice by its oral administration [5]. This activity was higher than that of clinically using anti-ulcer agent, sucralfate [5, 6]. Bupleuran 2IIc, which has a molecular weight of 63,000, was consisted of 85.8% of galacturonan region consisting of 70% of α(1->4) linked galacturonic acid and 30% of carboxymethylated galacturonic acid and branched galacturonic acid [4, 7]. Bupleuran 2IIc also contained ramified region which is consisted of rhamnogalacturonan core and several arabino and galactooligosaccharide side chains attached to either 2-linked rhamnosyl residue through 4-linked galacturonic acid or 2-linked rhamnose directly in the rhamnogalacturonan core [7, 8]. KDO-containing region, so-called rhamnogalacturonan II (RG-II) like region was also contained in bupleuran 2IIc as a minor region [7, 8]. Bupleuran 2IIc significantly inhibited a variety of gastric lesions in mice or rats [6]. The major mechanism of mucosal protection by bupleuran 2IIc was suggested due to its anti-secretary activity on acid and pepsin, its increased protective coating and its radical scavenging effects but not involved in the action of endogenous prostagrandins and mucus synthesis [6, 9].

Bupleuran 2IIc also stimulated blastgenesis (mitogenic activity) of spleen B lymphocytes in the absence of macrophages and T cells (Fig. 1), and led to antibody forming cells in the presence of IL-6 [11]. When pectin fraction, BR-2, which contains bupleuran 2IIc was administrated orally to C3H/HeJ mice for 7 consecutive days, proliferative responces of spleen cells were enhanced in the presence of bupleuran 2IIc, but another B cell mitogen, LPS did not give similar effect [10]. Among the structural moieties, ramified region was involved as the active site in the expression of the mitogenic

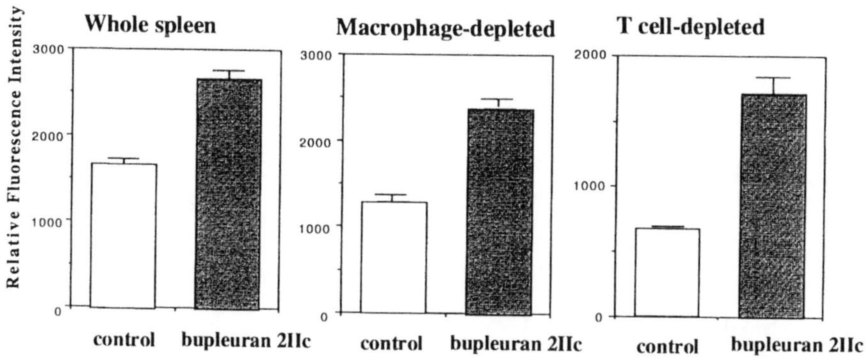


Fig. 1. Effect of adherent and T-cells in spleen cells on bupleuran 2IIc induced proliferation.

activity as similar as anti-complementary activity as reported in the previous study [4, 10]. Previous study also suggested that the ramified region and polygalacturonan region were involved in anti-ulcer activity of bupleuran 2IIc [5]. Therefore anti-polysaccharide polyclonal antibody against the ramified region from bupleuran 2IIc was made by rabbits, and applied to the studies on absorption and tissue distribution of bupleuran 2IIc [11]. When the pectin fraction (BR-2) from *B. falcatum* was administrated orally to the mice, the polysaccharide was detected in the liver and Peyer's patch by immunohistochemical staining and ELISA method using the anti-polysaccharide antibody [8]. The antigenic epitope against anti-polysaccharide antibody was characterized to be 6-linked galactosyl chains containing terminal glucuronic acid and 4-methylglucuronic acid which attached to (1->3)β-D-galactosyl chains by using specific carbohydrases(Fig. 2) [12]. Because mitogenic activity of bupleuran 2IIc was reduced in the presence of the antibody, the activity might be caused by same carbohydrate chains as the antigenic epitope. Therefore this antibody is expected to be useful for further studies on the molecular mechanism of the activity of bupleuran 2IIc.

## 3. Intestinal immune system modulating polysaccharides from rhizomes of *Atractylodes lancea*

Because traditional herbal medicines including Kampo medicines have generally been taken orally, some active ingredients may be absorbed from intestine, but some other active ingredients may affect Peyer's patch cells and intraepithelial lymphocyte (IEL), and follow by activation of mucosal immune systems in order to give several pharmacological acitvities. When

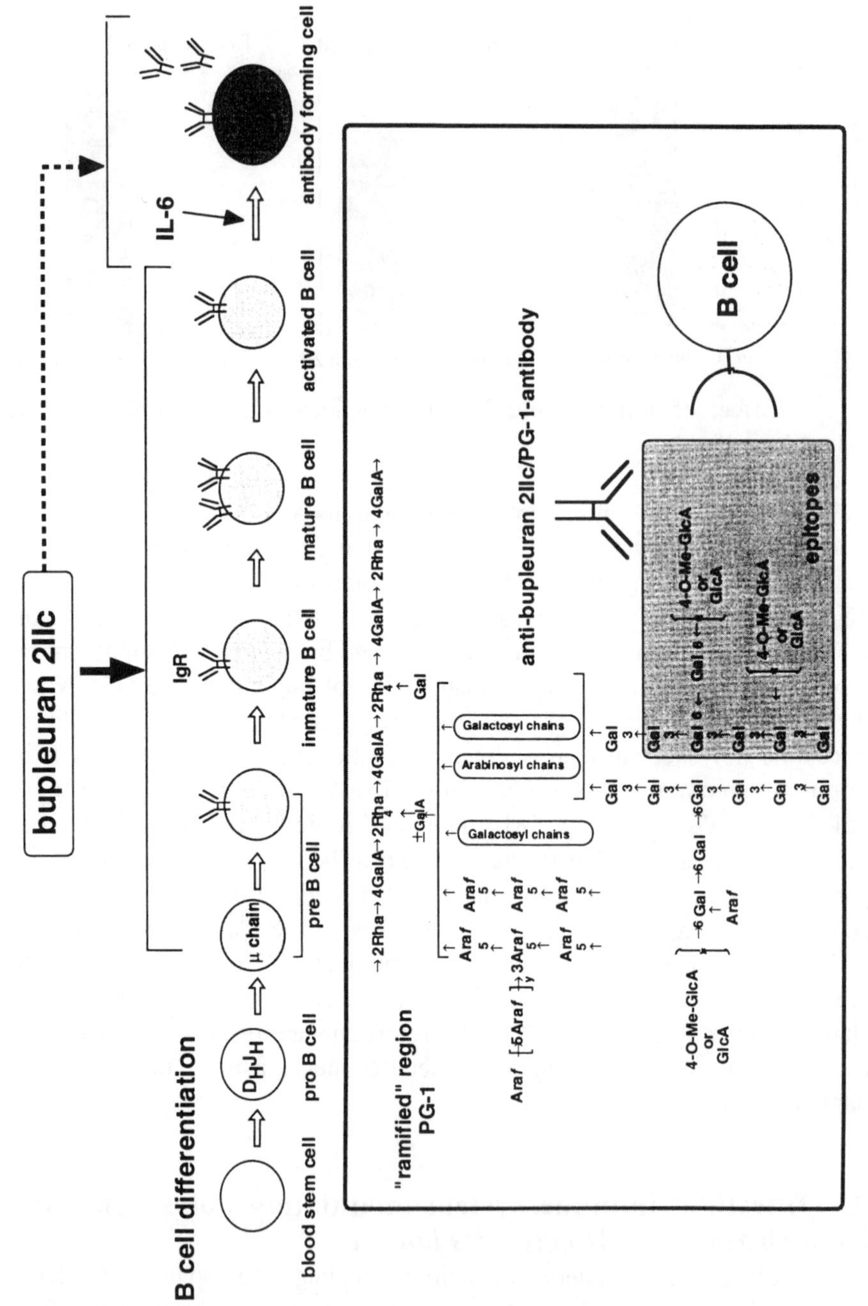


Fig. 2. Effect of bupleuran 2IIc on B cell differentiation and antigenic epitopes of bupleuran 2IIc/PG-1-antibody.

the cell suspension of Peyer's patch cells from the mice was cultured with polysaccharide fraction from rhizomes of *A. lancea*, the resulting culture supernatant enhanced the proliferation of bone marrow cells [13, 14]. This intestinal immune system modulating polysaccharide was purified, and the active polysaccharide, ALR-5IIa-1-1 was characterized to be arabino 3,6-galactan-containing polysaccharide [14]. ALR-5IIa-1-1 strongly reacted with β-glucosyl-Yariv antigen, suggesting the presence of arabino β-3,6-galactan moiety. Combination of exo-α-L-arabinofuranosidase and exo-β-D-(1→3)-galactanase digestions of ALR-5IIa-1-1 significantly reduced the activity (Fig. 3) [14]. These results suggest that arabino β-3,6-galactan moiety in the active polysaccharide contribute in the expression of intestinal immune system modulating activity.

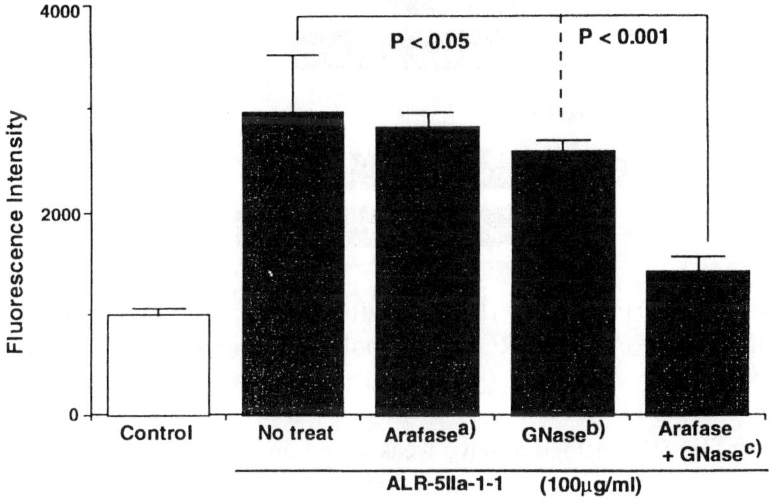


Fig. 3. Effect of enzymatic treatments on intestinal immune system modulating activity of ALR-5IIa-1-1

a) Exo-α–L-arabinofuranosidase digested
b) Exo-β–D-(1->3)galactanase digested
c) Exo-α–arabinofuranosidease and exo-β–galactanase digested

## 4. IL-6 production enhancing activity of rhamnogalacturonan II (RG-II) dimers from leaves of *Panax ginseng*

Although the roots of *Panax ginseng* is valuable because it takes 4-6 years for growing from the seed, the leaves of *Panax ginseng* can be harvested every year. Therefore, if the polysaccharides from the leaves have a similar activity as those from roots, the leaf polysaccharide will be available as well as the roots. Crude polysaccharide fraction from the leaves (GL-2) of *Panax ginseng* contained a large amount of TBA-positive substances suggesting the presence of Kdo and Dha which are characteristic component sugars of RG-II, but only little TBA-positive substance was contained in that from the roots (GR-2) [15]. Three RG-IIs (GL-4IIb-2, GL-5IIc and GL-5IId) were isolated from GL-2 in the leaves of *P. ginseng* without endo-polygalacturonidase digestion [15, 16]. Among them, GL-4IIb-2, which was originally isolated as macrophage Fc receptor expression enhancing polysaccharide [16, 17], had the most potent IL-6 production enhancing activity of macrophages (Fig. 4). GL-5IId was second potent one,

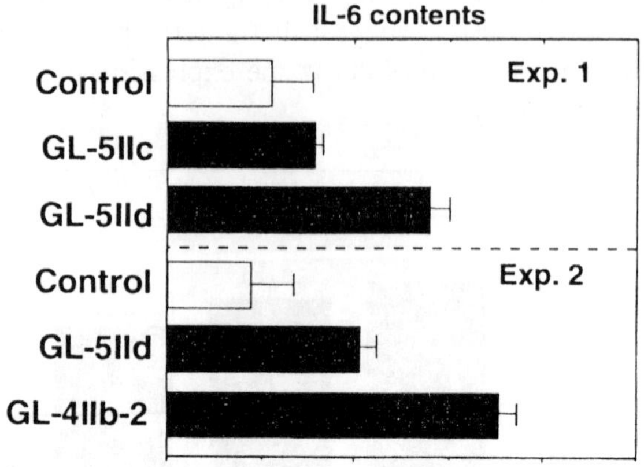


Fig.4 Comparison of IL-6 production enhancing activity of Rhamnogalacturonan II (RGII)-like polysaccharides from the leaves of *Panax ginseng*

however GL-5IIc seemed to have weak effect on IL-6 production. GL-4IIb-2 consisted of typical structural unit such as disaccharides of Rha-Kdo and Araf-Dha, AceA-containing oligosaccharide, and uronic acid-rich octasaccharide chains, and these oligosaccharides attached with oligogalacturonide core, as like observed in sycamore RG-II [16, 18]. But, AceA-containing oligosaccharide substituted another rhamnosyl side chain and composed nonasaccharide in GL-4IIb-2 (Fig. 5) [16].This was only difference from structural unit of sycamore RG-II. Although significant structural difference was not observed in three RG-IIs by the analyses of the oligosaccharide fragments, active RG-IIs, GL-4IIb-2 and GL-5IId were mainly present as a dimer cross-linked by borate diesters, but GL-5IIc was mainly of monomer [15]. Because dissociation of GL-5IId to the monomer decreased the activity, and the redimerization of the monomer recovered the activity, RG-II dimers may be responsible for expression of their activity (Fig 6) [15]. When we tested IL-6 production enhancing activity of another RG-II unit which was obtained from anti-ulcer pectin of *Bupleurum falcatum* by endo-polygalacturonase digestion, no activity was observed even when

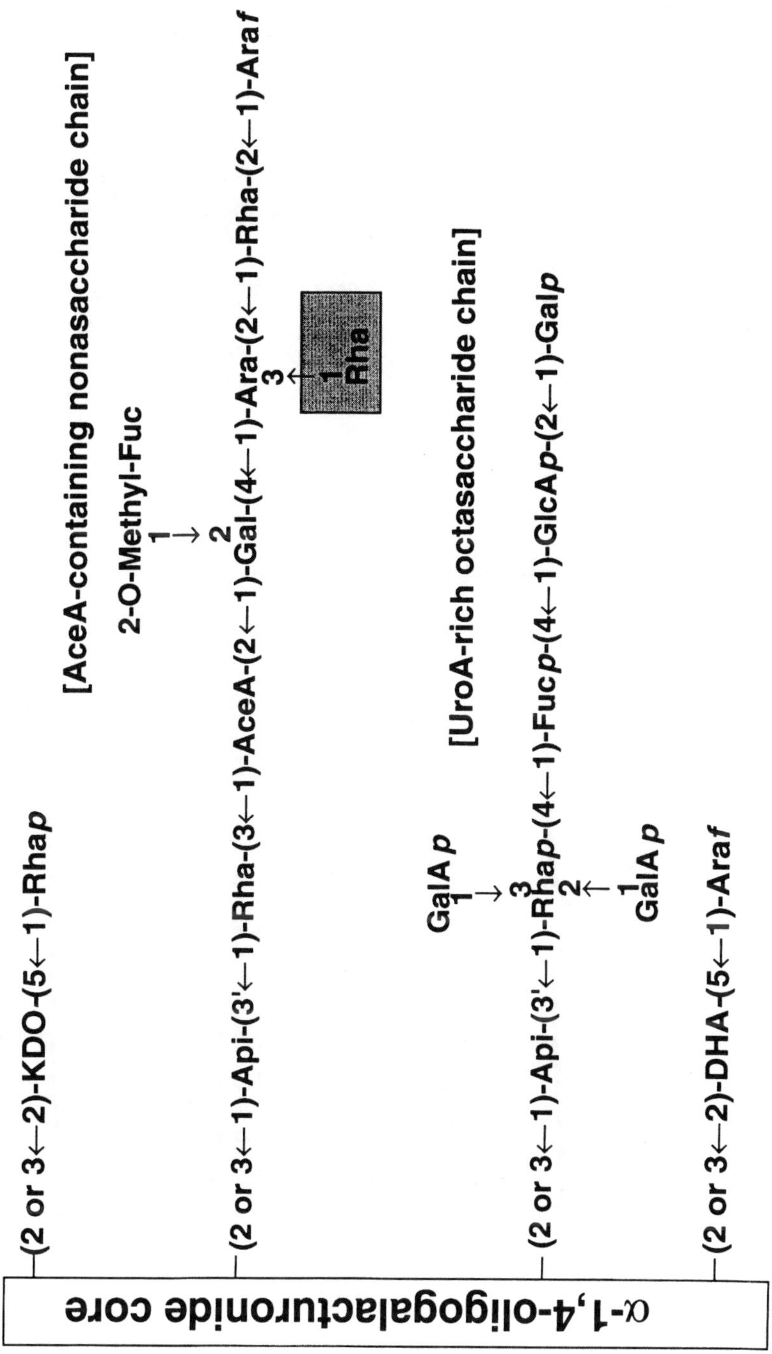

Fig. 5 Proposed structure of pharmacologically active RG-II (GL-4IIb-2) from the leaves of *Panax ginseng*

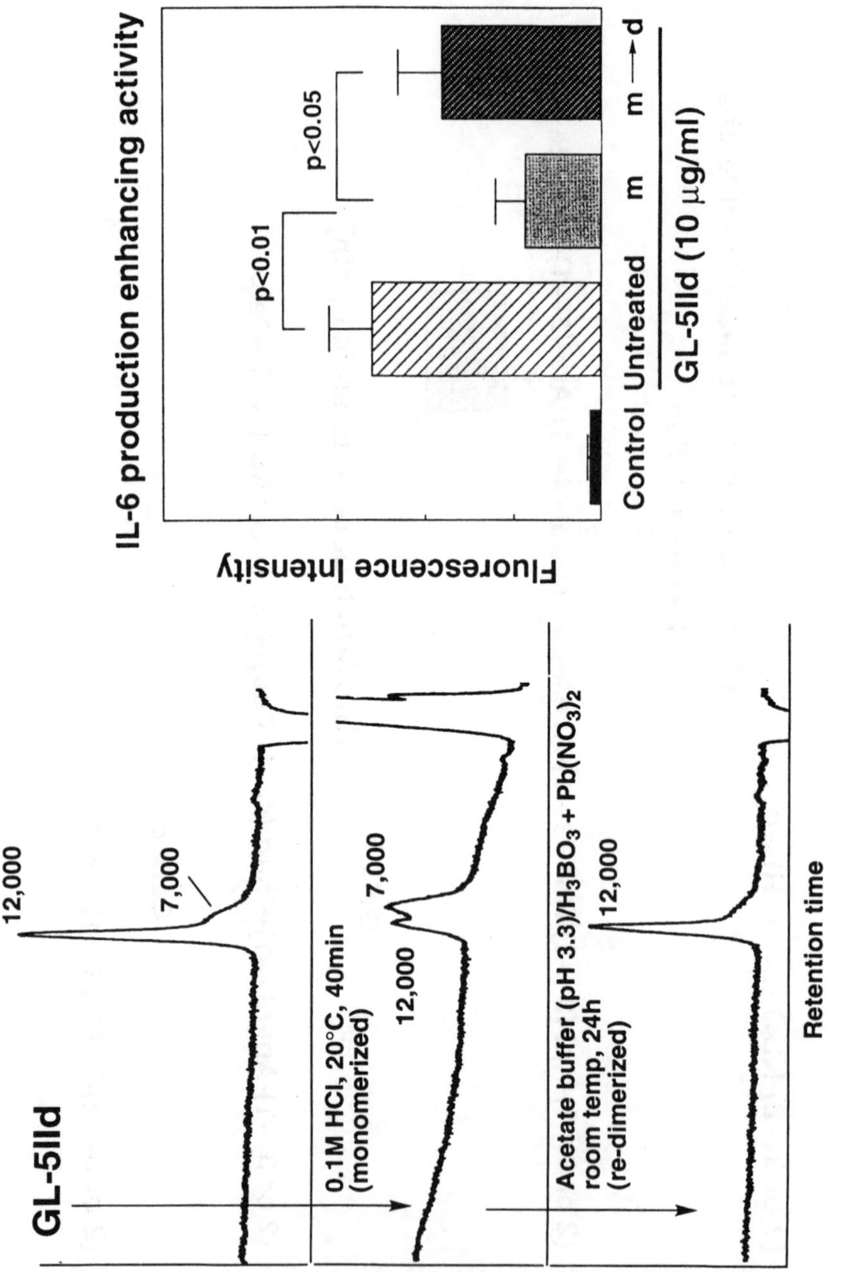


Fig. 6 Effect of boron complex formation on IL-6 production enhancing activity of GL-5IId; left: HPLC on Asahi-pak GS-320 + GS-220; right: IL-6 production enhancing activity

redimerized sample was used [15]. Therefore, these results suggest that the certain essential carbohydrate structure in RG-II may be necessary for expression of the activity although the dimerization contributes to expression of the potent activity.RG-II has generally been believed as one of the structural components in pectin since RG-IIs have been released from cell walls by endo-polygalacturonase digestion [19]. Because a relatively significant amount of free RG-II was contained in the hot water extract of ginseng leaves, the leaves may contain endo-polygalacturonase like enzyme in order to release RG-II like polysaccharides. In fact, significant endo-polygalacturonase activity was detected in the ginseng leaves but not roots [20]. Recently, we also found that RG-II dimer, GL-4IIb-2 from the leaves of *Panax ginseng*, also showed potent secretion enhancing activity of nerve growth factor (NGF) which is known to play as neurotrophic factor for survival of neuronal cells and prevention of aging and dementia (unpublished). These results expects that unique certain RG-II like polysaccharides may have several important biological activities.

## 5. Conclusion

Structural components of pectic polysaccharides such as ramified region, arabino 3,6-galactan and RG-II revealed variety of biological activities which depend on the difference of carbohydrate structures. These studies may give evidence for the efficacy of herbal medicine and new candidates for the development of carbohydrate originated medicine.

### References

[1] H. Yamada, (1994). Asia Pacific J. Pharmacol., 9, 209-217

[2] H. Yamada, (1994). Carbohydr. Polymers, 25, 269-276

[3] H. Yamada, (1996). Pectin and Pectinase, Progress in Biotechnology, Vol. 4 (Eds. J. Visser and A.G.J. Voragen), pp. 173-190, Elsevier, Amsterdam

[4] H. Yamada, K-S. Ra, H. Kiyohara, J-C. Cyong and Y. Otsuka, (1989). Carbohydr. Res., 189, 209-226

[5] H. Yamada, X-B. Sun, T. Matsumoto, K-S. Ra, M. Hirano and H. Kiyohara, (1991). Planta Medica, 57, 555-559

[6] X-B. Sun, T. Matsumoto and H. Yamada, (1991). J. Pharmacy Pharmacol., 43, 699-704

[7] H. Yamada, M. Hirano and H. Kiyohara, (1991).Carbohydr. Res., 219, 173-192

[8] M. Hirano, H. Kiyohara, T. Matsumoto and H. Yamada, (1994). Carbohydr. Res., 251, 145-162

[9] T. Matsumoto, R. Moriguchi and H. Yamada, (1993). J. Pharmacy Pharmacol., 45, 535-539

[10] M. Sakurai, T. Matsumoto, H. Kiyohara and H. Yamada, (1998) Abstracaat of 118th Annual Meeting of the Pharmaceutical Society of Japan, p. 134

[11] M. H. Sakurai, T. Matsumoto, H. Kiyohara and H. Yamada, (1996). Planta Med., 62, 341-346

[12] M.H. Sakurai, H. Kiyohara, T. Matsumoto, Y. Tsumuraya, Y. Hashimoto and H. Yamada, Carbohydr. Res., in press.

[13] T. Hong, T. Matsumoto, H. Kiyohara and H. Yamada, Phytomedicine, in press.

[14] K-W. Yu, H. Kiyohara, T. Matsumoto, H-C. Yang and H. Yamada, Planta Med., in press.

[15] K-S. Shin, H. Kiyohara, T. Matsumoto and H. Yamada, (1998). Carbohydr. Res., 307, 97-106

[16] K-S. Shin, H. Kiyohara, T. Matsumoto and H. Yamada, (1997). Carbohydr. Res., 300, 239-249

[17] X-B. Sun, T. Matsumoto and H. Yamada, (1994). Phytomedicine, 1, 225-231

[18] A.J. Whitcombe, M.A. O'Neill, W. Steffan, P. Albersheim and A.G. Darvill, (1995). Carbohydr. Res., 271, 15-29

[19] P. Albersheim, J. An, G. Freshour, M.S. Fuller, R. Guillen, K-S. Ham, M.G. Hahn, J. Huang, M.A. O'Neill, A.J. Whitcombe, M.U. Williams, W.S. York and A.G. Darvill, (1994). Biochem. Soc. Trans., 22, 374-378

[20] K-W. Yu, H. Kiyohara, M. Sakurai and H. Yamada, (1998). Abstract of 20th Japanese Carbohydrate Symposium, p. 125

# Chapter 3

# Bioactive compounds from Iceland moss

K. INGÓLFSDÓTTIR
*Department of Pharmacy, University of Iceland, Reykjavik, Iceland*

Key words: Iceland moss, *Cetraria islandica*, lichen, polysaccharides, immunomodulating

Abstract: Pharmacological investigations of the lichen *Cetraria islandica* (L.) Ach. have shown that polysaccharides as well as low molecular weight constituents exhibit significant biological activity. A polysaccharide with a backbone of (1→6)-linked α-D-mannopyranosyl and α-D-(1→6)-galactopyranosyl units has been isolated from an alkali extract of Iceland moss. The galactomannan (mean $M_r$ = 18 kdalton) exhibited pronounced enhancement of phagocytosis in both *in vitro* and *in vivo* assays. In addition, several polysaccharide fractions, isolated from a hot aqueous extract of Iceland moss by ethanol fractionation and ion-exhange chromatography, exerted significant activity in several immunological assays. The fractions showed *in vitro* anti-complementary activity, pronounced enhancement of *in vitro* granulocytic phagocytosis and a significant increase in the rate of colloidal carbon elimination in the *in vivo* carbon clearance test. An α-(1→3)-(1→4)-glucan (mean $M_r$ = 2,000 kdalton) has been isolated in pure form from two of the aqueous fractions.
Of low molecular weight constituents, the aliphatic α-methylene-γ-lactone (+)-protolichesterinic acid has exhibited prominent activity in several *in vitro* biological assays. With reference to the traditional use of Iceland moss, inhibitory effects of plant constituents on arachidonate metabolism and *Helicobacter pylori* have been studied. In both cases activity was detected and attributed to protolichesterinic acid, which exhibited dose-dependent inhibitory



*B.S. Paulsen (ed.), Bioactive Carbohydrate Polymers,* 25–36.

effects ($IC_{50}$ = 20 μM) on the enzyme 5-lipoxygenase and an MIC range of 16-64 μg/ml (n = 35) against *H. pylori*. Protolichesterinic acid has furthermore been shown to have marked anti-proliferative activity against two breast cancer cell lines (T-47D, ZR-75-1) and the leukemia cell line K-562 ($ED_{50}$ = 3-15 μM) without affecting normal lymphocytes and human skin fibroblasts .

## 1. Introduction

Despite being called moss, Iceland moss (*Cetraria islandica*) is botanically a lichen, *i.e.* it is formed through a symbiotic association between a fungus and an alga.

Although most lichens are quite small plants, especially the crustose lichens that grow on rocks, a number of species have been used medicinally throughout the world, *e.g.* in Europe, China, South America. Lichens that have been used medicinally are usually species that can be collected in reasonable amounts without too much effort. *Peltigera aphthosa*, for example, was used medicinally in Norway for treating eye infections and skin disease characterized by formation of blisters. Indications for the use of lichens as a group of plants for medicinal purposes are very wide-ranging.

It is not known why the name of the lichen *Cetraria islandica* is referred to Iceland; the plant grows in many other places and has been used in folk medicine of many countries. Medicinally the plant has been used for a number of conditions ranging from tuberculosis to throat irritation, in addition to being used to promote general health. In Iceland the plant is furthermore used as a food. Iceland moss is one of a very few lichen species used medicinally in Europe today. The medical indications accepted by European health regulatory authorities for human use of Iceland moss are dry cough and irritation of the upper respiratory tract, laryngitis, and lack of appetite. There have been no reports of drug interactions or toxicity arising from the use of Iceland moss.

We have conducted chemical and pharmacological studies of Iceland moss with the aim of finding answers to two key questions: 1) Can claimed benefits of Iceland moss be substantiated by scientific means? 2) Can lead compounds for drug discovery be found in this plant?

In choosing biological assays, reference has been made to traditional use, but the plant has also been screened for activity not linked to traditional use. Results have shown that bioactive compounds can be found amongst low molecular weight compounds as well as high molecular weight, *ie.* polysaccharides.

This presentation will focus on bioactive polysaccharides, most notably those exhibiting immunomodulating activity. Towards the end of the presentation, referral will be made to bioactive compounds of low molecular weight.

The best known polysaccharides from previous chemical studies of *C. islandica* are lichenan and isolichenan. Both are glucans, lichenan containing ß-(1→3) and ß-(1→4) linkages and isolichenan containing α-(1→3) and α-(1→4) linkages. These glucans have been known for quite some time. More recently a Yugoslavian research group reported the isolation of an alkali-soluble polysaccharide containing D-glucose and D-glucuronic acid units (1). In 1984 a branched galactomannan was isolated from a hot water extract and obtained through an insoluble copper complex and contained mannose, galactose and glucose in a molar ratio 43:40:17 (2). The presence of glucose was attributed to lichenan as an impurity.

Several studies have been performed previously on the biological activity of crude polysaccharide fractions from *Cetraria*. Investigations have for example shown that a crude polysaccharide fraction from a Japanese variety of *C. islandica* (var. *orientalis* Asahina) exerted significant growth inhibitory effects on sarcoma-180 tumours in mice (3, 4). The polysaccharides were shown to be non-cytotoxic the activity was suggested to be host-mediated.

When the present investigation began, purified polysaccharides from Iceland moss had not been screened for biological activity nor had purified polysaccharides from lichens in general been investigated for immunomodulating properties.

## 2. Isolation of polysaccharides

For the isolation of polysaccharides from Iceland moss, the plant material was first extracted with organic solvents to remove small molecules, followed by hot water (Fig.1). Lichenan was removed by the traditional method of freezing and thawing the aqueous extract. This glucan is soluble in hot water, but forms a rigid gel on cooling. The extract was then fractionally precipitated with ethanol and the precipitates separated into neutral and acidic fractions through ion-exchange chromatography. By the combined use of UV-, refractometric- and polarimetric detetection, 10 fractions were obtained.

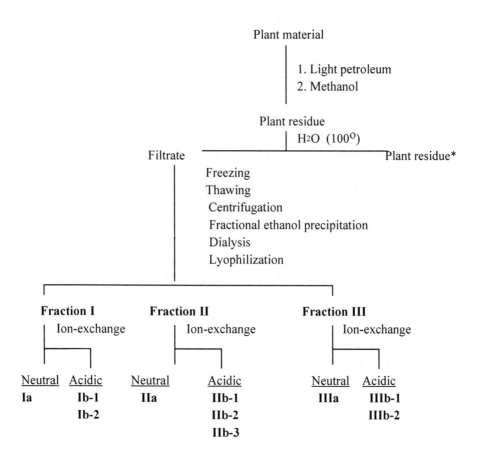

*Fig. 1.* Fractionation of polysaccharides from aqueous extract of *Cetraria islandica.*

The plant residue from the water extraction was subsequently extracted with alkali (Fig. 2) and the extract treated similarly to the water extract, *ie.* by fractional ethanol precipitation. Protein was precipitated with TCA (trichloroacetic acid). Four neutral and acidic fractions were obtained.

The mean $M_r$ of the major components of the polysaccharide fractions, determined by high pressure gel permeation chromatography (HP-GPC), ranged from approximately 18 kdalton to 2000 kdalton. Of the total number of fractions, 5 were obtained in quite small quantity (<0.01% dry weight) and not pursued further.

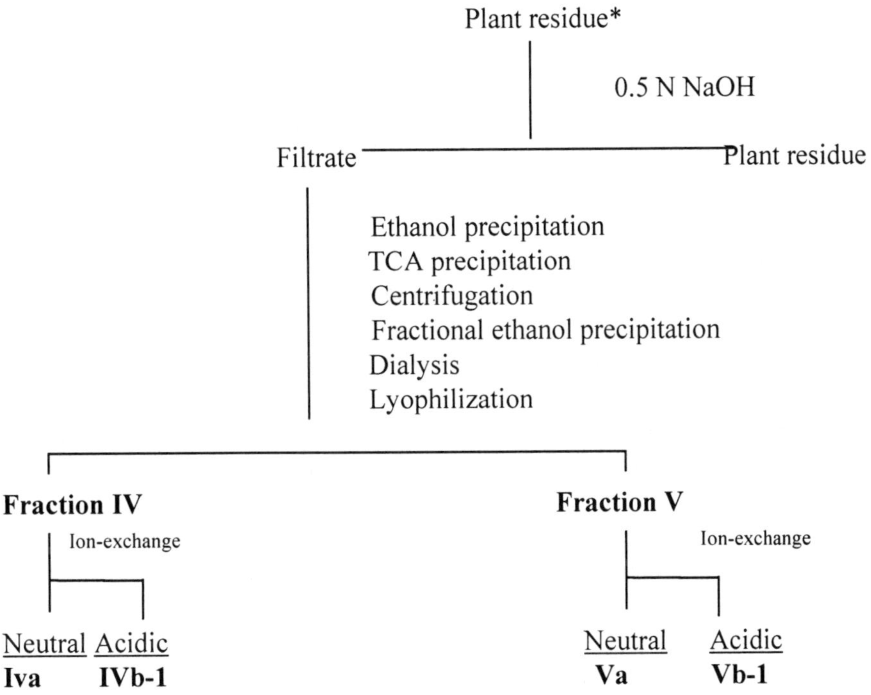


*Fig. 2.* Fractionation of polysaccharides from alkali extract of *Cetraria islandica.*

*Table 1.* *In vitro* phagocytosis assay of chief acidic and neutral polysaccharide fractions from aqueous and alkali extracts of *Cetraria islandica.* Results are expressed as % relative stimulation as compared with control, set as 100% (fMLP, 1.0 mg/ml). Fractions are designated as in Figs. 1 and 2. (-) Inactive

| **Fraction** | | **Concentration** | **(μg/ml)** | | |
|---|---|---|---|---|---|
| | **1000** | **100** | **10** | **1** | **0.1** |
| I-a | 48.9 | 53.9 | 18.1 | 5.0 | 4.0 |
| II-a | 63.5 | 50.5 | 23.1 | 8.2 | -1.7 |
| II-b-1 | 55.1 | 38.3 | 21.5 | 11.5 | 6.0 |
| III-a | (-) | (-) | (-) | (-) | (-) |
| III-b-1 | 55.4 | 24.3 | 6.1 | -4.1 | -4.4 |
| IV-a | 62.2 | 29.7 | 9.0 | 6.8 | 6.6 |
| IV-b-1 | (-) | (-) | (-) | (-) | (-) |
| V-a | (-) | (-) | (-) | (-) | (-) |
| V-b-1 | 64.5 | 38.6 | 12.1 | 8.4 | -2.1 |

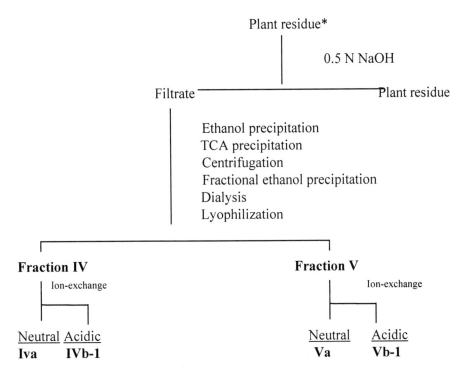

Each of the major fractions was screened for potential immunostimulating activity by submitting them to an *in vitro* phagocytosis assay performed with human granulocytes (Table 1). Of the 9 fractions tested, 6 showed significant enhancement of granulocytic phagocytosis. One of the most active fractions was V-b-1 from the alkali extract and this fraction became the first choice for further purification.

## 3. Purification of fraction V-b-1 to give KI-M-7

The fraction V-b-1 was purified through gel filtration using Sephacryl S-400 HR. For detection, a combination of UV spectroscopy, refractometry and polarimetry was used.

The molecular weight of the purified polysaccharide, which was given the name KI-M-7, proved to be approximately 18 kdalton as determined by HP-GPC. The neutral sugar components were identified by GC analysis as galactose, mannose and rhamnose in a molar ratio of 13:9:1. The absence of uronic acid residues was confirmed through the carbazole test and NMR data.

The structure of the heteroglycan proved to be highly complex. Based on methylation analysis, partial hydrolysis and $^{13}$C NMR data, it is proposed that the polysaccharide is composed of two carbohydrate blocks and that the minimal repeating unit of the polysaccharide is composed of 10 types of component sugar units (5). The core of one block contains α-D-(1→6)-linked mannopyranose units, a portion of which are either 2-*O*- or 4-*O*-substituted. The core of the other block contains α-D-(1→6)-galactopyranosyl units which are either unsubstituted or 2,4-di-*O*-substituted. Terminal units are composed of α-D-galactopyranosyl and β-D-galactofuranosyl groups.

When subjected to *in vitro* phagocytic testing, KI-M-7 proved quite active, showing an increase of granulocytic phagocytosis amounting to 68% at a concentration of 100 μg/ml (Table 2). This is in correlation with the result for the initial crude fraction, V-I-b, which showed 38% stimulation when tested at the same concentration.

*Table 2. In vitro* phagocytosis assay of purified polysaccharide KI-M-7 from fracion V-b-1 and purified polysaccharide Ci-3 from fractions Ia, IIa and IV-a of *Cetraria islandica*. Results (mean ± SD from 3 experiments) are expressed as % relative stimulation as compared with control, set as 100% (fMLP, 1.0 mg/ml).

| Polysaccharide | Concentration | µg/ml | |
|---|---|---|---|
| | 100 | 10 | 1 |
| KI-M-7 | 68.0 ± 4.9 | 24.0 ± 2.4 | -14.0 ± 6.2 |
| Ci-3 | 46.0 ± 2.2 | 24.0 ± 2.3 | 6.0 ± 7.8 |

As a follow-up, KI-M-7 was tested in the carbon clearance test, which is an *in vivo* phagocytosis assay. In this assay, phagocytic activity is determined by the rate of removal of injected colloidal carbon particles from the blood of mice. Test substances are administered *i.p.* to the mouse and the rate of removal of the foreign particles is compared between controls (n=8) and the test group (n=8). In this case, the polysaccharide gave rise to an increase in the rate of carbon elimination by a mean ratio of 1.9 compared with controls, which represents high activity (5). The results reflect a marked increase in the rate of phagocytosis and suggest significant activation of the reticuloendothelial system (RES).

The result of the carbon clearance assay for this lichen polysaccharide is comparable to results for the fungal ß-glucan lentinan, which is used clinically in Japan.

## 4. Immunological testing of aqueous polysaccharides

Following purification of KI-M-7 from the alkali extract, attention was focused on the water soluble polysaccharides since these would be likely constituents of traditional preparations. Phagocytic results for the water-soluble polysaccharide fractions showed several fractions to be active (6). At the highest concentration (1000 µg/ml), stimulation of granulocytic phagocytosis amounting to 49-63% was exhibited by 4 of the 5 major fractions as compared to control, set as 100% (Table 1). The most active fraction was IIa.

In addition to the phagocytic assay, the major aqueous fractions were also subjected to anti-complementary screening (Table 3). Over-activation of the complement system has been shown to contribute to the etiology of many inflammatory diseases, including rheumatoid arthritis, asthma, adult respiratory distress syndrome and systemic lupus erythematosus (7). Compounds exhibiting anti-complementary effects would therefore be of potential interest as aids in the treatment of such diseases. Even at the lower concentration (100 µg/ml) all fractions showed considerable reduction of complementary-induced hemolysis, ranging from 49-85% (6).   Most active fraction again was IIa.

*Table 3. In vitro* anti-complementary assay (classical pathway) of chief acidic and neutral polysaccharide fractions from aqueous extract of *Cetraria islandica*.   Results (means ± SD from 3 experiments) are expressed as % reduction of complementary-induced hemolysis.

| Fraction | Concentration (µg/ml) | |
|---|---|---|
| | 1000 | 100 |
| I-a | 85.5 ± 4.7 | 68.5 ± 23.5 |
| II-a | 90.2 ± 5.1 | 85.1 ± 2.0 |
| II-b-1 | 102.5 ± 1.9 | 63.7 ± 13.0 |
| III-a | 88.0 ± 4.0 | 69.9 ± 2.8 |
| III-b-1 | 79.5 ± 4.1 | 49.3 ± 13.6 |

## 5.   Purification of fractions IIa, Ia and IV-a to give Ci-3

From the phagocytic and anti-complementary results, fraction IIa was chosen for further purification using gel filtration.   Through methanolysis, methylation analysis, optical rotation and NMR spectroscopy, the identity of the main polysaccharide has been confirmed as a glucan containing α-(1→3) and α-(1→4) linkages in a ratio of 2:1 (8).   The structure thus resembles that of isolichenan, but the present polysaccharide has a much higher degree of polymerization, the molecular weight being 2000 kdalton instead of 6-8 kdalton. It has become clear that Ci-3 is the major polysaccharide in two additional fractions:   Ia from the water extract (Fig. 1) and IV-a from the alkali extract (Fig. 2).

Results for phagocytic testing of Ci-3 *in vitro* show that the polysaccharide exerts 46% stimulation at 100 µg/ml (Table 2), and is thus a little less active than KI-M-7.   It is currently undergoing *in vivo* phagocytic investigations in the carbon clearance assay.   Potent anti-complementary activity of Ci-3 has been confirmed, a 90% reduction of complementary-induced hemolysis being expressed at a concentration of 100 µg/ml (8).

## 6.   Traditional extract

The traditional and current method of using Iceland moss prior to consumption is by boiling the plant directly with water (*vs.* removal of low molecular weight compounds) and for shorter lengths of time, 10-15 minutes *vs.* 2 hours.   In order to establish whether the traditional extraction method was sufficient to bring out immunomodulating polysaccharides, a traditional extract was prepared and subjected to phagocytic testing (6).   In the *in vitro* phagocytosis assay, the lyophilized traditional extract showed enhancement of granulocytic phagocytosis amounting to 92.6% at a concentration of 100 µg/ml (Table 4).

*Table 4. In vitro* phagocytosis assay of traditionally prepared extract of *Cetraria islandica*; 1) lyophilized total extract, 2) polysaccharide fraction, 3) low molecular weight fraction. Results (mean ± SD from 3 experiments) are expressed as % relative stimulation as compared with control, set as 100% (fMLP, 1.0 mg/ml).

| Traditional extract | Concentration | (µg/ml) | |
| --- | --- | --- | --- |
| | 100 | 10 | 1 |
| Total extract | 92.6 | 34.8 | |
| Polysaccharide fraction | -25.4 | 46.9 | 6.5 |
| Low MW fraction | 7.6 | -4.2 | 2.4 |

In order to establish whether the effects were solely due to the polysaccharide content, an extract prepared in the same way was fractionated into low molecular weight- and polysaccharide fractions and these tested as before. Results indicate that the phagocytic activity of the traditional extract was mainly due to the presence of polysaccharides (Table 4).

The traditionally prepared extract, untreated as well as fractionated, was further subjected to *in vivo* phagocytic testing. The untreated extract stimulated the rate of carbon elimination by a mean ratio of 1.9 compared with controls (6). This degree of stimulation is taken to reflect a significant increase in the rate of phagocytosis and suggests prominent activation of the reticuloendothelial system. Results for the fractionated extract showed that the low molecular weight fraction contributes significantly to the *in vivo* phagocytic results, increasing the rate of carbon elimination by a mean ratio of 1.5. This is in contrast to results of the *in vitro* phagocytosis assay, and could indicate that *in vivo* stimulation of low molecular weight compounds is lymphocyte- or interleukine- associated.

In summary, the work presented has led to the isolation and purification of two new polysaccharides from Iceland moss, both of which exhibit significant immunomodulating properties. It is further clear that there are other active polysaccharides present in the plant. At least some of the immunomodulating polysaccharides are obtainable through relatively brief extraction, *cf.* traditional methods.

The results in our opinion definitely warrant continued investigations. For one thing, the purification and structure elucidation of the remaining active polysaccharide fractions need to be carried out. The implications of immunomodulating activity of low molecular weight compounds should also be followed up. Further *in vivo* studies should be done and ultimately the question of whether plant constituents are active in the human body should be addressed.

## 7. Bioactive low molecular weight compounds

In addition to immunological assays, crude extracts of Iceland moss have been tested in various *in vitro* biological models. In these instances, where activity has been found, it has been due to low molecular compounds, and most particularly to the aliphatic α-methylene lactone (+)-protolichesterinic acid (Fig. 2).

With reference to the traditional use of Iceland moss to treat bronchial and inflammatory conditions, the effects of crude extracts on two key enzymes involved in arachidonate metabolism, *ie.* 5-lipoxygenase and cyclooxygenase, were tested for. Protolichesterinic acid was identified as an active constituent, selectively inhibiting the enzyme 5-lipoxygenase *in vitro* in a dose dependent manner ($IC_{50}$= 20μM) without significantly inhibiting cyclooxygenase (9). Products of the lipoxygenase pathway are mediators of bronchoconstriction and inflammatory responses, so inhibition of these enzymes could theoretically lead to relaxation of bronchial muscle and antiinflammatory effects.

*Cetraria islandica* has been used in Iceland to relieve symptoms of gastric- and duodenal ulcer. Extracts of the lichen were therefore tested for *in vitro* inhibitory activity against *Helicobacter pylori*, the organism reputed to contribute to the etiology of gastric- and duodenal ulcer as well as to that of gastric cancer.

Inhibitory activity was exhibited by the petroleum ether extract. The active compound was identified as (+)-protolichesterinic acid (10). The MIC for protolichesterinic acid was determined using 35 strains of *Helicobacter*, mainly clinical isolates but a standard strain was also used. The MIC range proved to be 16-64 μg/ml (10).

The MIC of protolichesterinic acid is considerably higher than that of drugs currently used in the eradication of *H. pylori*, but it is tempting to speculate whether the reputed beneficial effects of Iceland moss in cases of gastritis and gastric/duodenal ulcer could be due in part to inhibitory activity against *H. pylori.*

The effects of protolichesterinic acid have further been established on livng human cells in culture. For these investigations 3 malignant cell lines (breast cancer cell lines T-47D and ZR-75-1, erythro-leukemia cell line K-562) and 2 normal cell lines (skin fibroblasts, peripheral blood lymphocytes) were chosen.

The effects of protolichesterinic acid on DNA synthesis in malignant cells were tested by measuring $^3$H-thymidine uptake. Protolichesterinic acid proved to have marked anti-proliferative effects on both breast cancer cell lines, with $ED_{50}$ values of 3.8 µg/ml for T-47D and 1.1 µg/ml for ZR-75-1 (11). Protolichesterinic acid also had marked inhibitory effects on $^3$H-thymidine uptake in the leukemic cell line K-562 with an $ED_{50}$ value of 11.2 µg/ml (11).

Protolichesterinic acid had neither anti-proliferative nor morphological effects on normal human skin fibroblasts and resting peripheral blood lymphocytes in culture (11).

In summary, a lot of work remains to be done before answers can be given to the two questions set out with, *ie.* whether claimed benefits of Iceland moss can be confirmed by scientific investigations and whether lead compounds for drug discovery can be found in this plant. However, from the knowledge acquired so far, both with regard to low molecular- and high molecular weight compounds, it seems that the quest should be continued.

**Acknowledgements**
The results presented have been acquired through extensive collaboration between a number of institutions and individuals, whom I would like to thank and whose names are cited in the references.

## References

[1] Hranisavljevic-Jakovljevic M., Miljkovic-Stojanovic J., Dimitrijevic R., Micovic V.M. (1980).Carbohydr. Res. 80, 291-295.
[2] Gorin P.A. and Iacomini M. (1984).Carbohydr. Res. 128, 119-132.
[3] Shibata, S., Nishikawa, Y., Tanaka, M. (1968). Z. Krebsforsch. 71: 102-104.
[4] Fukuoka, F., Nakanishi, M., Shibata, S., Nishikawa, Y., Takeda, T., Tanaka, M. (1968). Jap. J. Cancer Res. 59: 421-432.
[5] Ingólfsdóttir, K., Jurcic, K., Fischer, B. and Wagner, H. (1994). 60(6), 527-531.
[6] Ingólfsdóttir K., Jurcic K., Wagner H.(1998) Phytomedicine. In press.
[7] Morgan, B.P. (1996) Biochem. Soc. Trans. 24(1): 224-229.

[8] Ólafsdóttir E.S., Ingólfsdóttir K., Barsett H., Smestad-Paulsen B., Jurcic K., Wagner H.(1998) Immunologically activ (1→3)-(1→4)-α-D-glucan from Cetraria islandica. Proceedings: Bioactive Carbohydrate Polymers, Phytochemical Society of Europe, Oslo.

[9] Ingólfsdóttir, K., Breu, W., Huneck, S., Gudjonsdottir, G.A. and Wagner, H. (1994).. Phytomedicine 1, 187-191.

[10] Ingólfsdóttir K., Hjálmarsdóttir M.Á.,Gudjónsdóttir G.A., Brynjolfsdottir A., Sigurdsson A., Steingrímsson Ó. (1997). Antimicrob. Agents Chemother. 41(1), 215-217.

[11] Ögmundsdóttir H., Zoëga G. M., Gissurarson S. R., Ingólfsdóttir K. (1998) J. Pharm. Pharmacol. 50(1), 107-115.

Chapter 4

# Structural features of biologically active polysaccharide fractions from the leaves and seeds of *Plantago major* L.

A.B.SAMUELSEN
*Department of Pharmacognosy, School of Pharmacy, University of Oslo, P.O.Box 1068 Blindern, N-0316 Oslo, Norway*

Key words: *Plantago major* L., polysaccharides, pectin, arabinogalactan, heteroxylan, anti-complementary activity

Abstract: The leaves of *Plantago major* L. have been used as a wound healing remedy for centuries in many parts of the world. Polysaccharide fractions have been isolated from the leaves and screened for biological activity using the test for anti-complementary activity [1]. Fractions PMIa and PMII which were isolated from the 50(C water extract had high anti-complementary activity, and both fractions were subjected to structural analysis. PMIa was an arabinogalactan type II [2] and PMII was a pectin that contained both smooth and hairy regions [3-5]. Structure-activity studies showed that the ramified region PVa was the part of PMII that had the highest anti-complementary activity [4]. Polysaccharide fractions that contained acidic heteroxylans with anti-complementary activity have been isolated from the seeds [6]. The heteroxylans contained xylan backbones that were composed of blocks of (1→3)-linked residues and blocks of (1→4)-linked residues. Short side chains of single arabinose and xylose residues, Ara*f*-(1→3)-Xyl*p* and Glc*p*A-(1→3)-Ara*f* were linked to *O*-2 or *O*-3 of (1→4)-linked Xyl*p* residues in the backbone [6, 7].



*B.S. Paulsen (ed.), Bioactive Carbohydrate Polymers,* 37–46.

# 1. Introduction

*Plantago major* L. is a perennial plant that is widely spread throughout the world. It has many common names in English: large plantain, common plantain or great plantain. In Norway it is called *groblad,* which means «healing leaves» or «growing leaves». That is because the leaves are used in traditional medicine as a wound healing remedy.

The seeds of *P. major* contain polysaccharides in the seed coat which swell in contact with water. This make the seeds sticky, and they can easily get attached to humans or animals and thereby be spread. *P. major* was spread by man from Europe and throughout the world. The Indians named it «White man's footprint» because it was found everywhere the Europeans had been. This was later adapted into the genus name *Plantago* which is from Latin *planta,* meaning sole of the foot.

*P. major* is an old medicinal plant. It has been used as a wound healing remedy for at least 2000 years and was described by Dioscorides in "De materia medica" in the 1st century [8]. From literature, it is known that the Vikings used it [9] and that it was commonly used in the time of Shakespeare. *P. major* is mentioned in «Romeo and Juliet» written by Shakespeare in the period 1592-1609 (Act I, Scene II):

*Romeo: Your plantain leaf is excellent for that*
*Benvoleo: For what, I pray thee?*
*Romeo: For your broken shin.*

*P. major* leaves are used for wound healing in many parts of the world; in Asian, European and American countries [10-15]. Either whole or crushed leaves are applied directly on wounds. Some places water extracts are used.

Biologically active polysaccharide fractions isolated from *P. major* have been studied. Polysaccharide fractions were isolated from the leaves and screened for biological activity using the test of anti-complementary activity [16]. Structural analysis was performed on the most active fractions. Polysaccharides were also isolated from the seeds for testing of anti-complementary activity and structure elucidation. In this chapter the main results from these investigations are summarised.

## 2. Isolation of leaf polysaccharide fractions

The leaves were pre-extracted with 80% ethanol at 80°C for removal of low-molecular weight compounds. Three crude polysaccharide extracts were obtained from the remaining leaf material after sequential water extraction at 50°C and 100°C and finally with DMSO. Each of the crude extracts was separated into one neutral (PMN) and two acidic fractions (PMI and PMII) by ion-exchange chromatography (IEC) using a DEAE-Sepharose fast flow column (Pharmacia) and a NaCl gradient (0-1M). The three crude extracts gave similar profiles [1], and the different PMN fractions, the PMI and PMII fractions had similar monosaccharide compositions as shown in Table 1.

**Table 1.** Monosaccharide composition (mol %) and protein content (w/w %) of polysaccharide fractions isolated from *Plantago major* L. leaves. Reproduced from [1] with permission from John Wiley & Sons Limited, Copyright & Licensing Department.

| | PMN | | | PMI | | | PMII | | |
|---|---|---|---|---|---|---|---|---|---|
| | 50 | 100 | DMSO | 50 | 100 | DMSO | 50 | 100 | DMSO |
| Ara | 6,4 | 9,2 | 35,1 | 23,9 | 12,4 | 22 | 5,9 | 4,1 | 5,5 |
| Rha | - | - | - | 3,1 | 2,2 | 3 | 3,3 | 2,6 | 3,2 |
| Xyl | 3,4 | 3,5 | - | 21,7 | 22,2 | 11,5 | - | - | - |
| Man | 28,1 | 36,3 | 11,7 | 2,8 | 1,2 | - | 1,7 | - | - |
| GalA | - | - | - | 5,9 | 38,6 | 22,9 | 74,8 | 86,2 | 81,3 |
| Gal | 8,4 | 8,9 | 11,9 | 34,1 | 18,1 | 28,2 | 8,1 | 4,3 | 6 |
| Glc | 53,8 | 42,2 | 41,3 | 8,6 | 5,5 | 12 | 6,3 | 2,8 | 4 |
| protein | - | - | 1 | 1,6 | 1,8 | 1,1 | 0,8 | 0,1 | - |

Despite of these similarities, only the acidic fractions PMI and PMII from the 50°C crude extract had high anti-complementary activity [1], and these fractions were subjected to structural analysis.

### *Arabinogalactan*

IEC fraction PMI from the 50°C leaf water extract was separated into two fractions, PMIa and PMIb by size exclusion chromatography (SEC) using a Sephacryl S-400 HR column (Pharmacia). PMIa had the highest anti-complementary activity of these two and was therefore subjected to further analysis [2].

According to its monosaccharide composition and to its positive reaction with Yariv β-glucosyl reagent, there were strong indications of the presence of an arabinogalactan type II [17] in PMIa. PMIa consisted of 93% carbohydrate and 1.5% protein, and the molecular weight was estimated to 77-80 kDa. The carbohydrate part consisted of 38% arabinose, 49% galactose, 6% rhamnose and 7% galacturonic acid. The protein part contained relatively high amounts of hydroxyproline (29%), alanine (15%) and serine (11%). PMIa constituted about 2% of the carbohydrates in the crude extract.

Methylation analysis showed that PMIa contained a heavily branched galactan: 50% of the Gal*p* residues were (1→3,6)-linked, about 10% were (1→3)-linked and 15% (1→6)-linked. The Ara*f* residues were mainly (1→5)- and terminally linked. Smith degradation of PMIa revealed the presence of a (1→3)-linked galactan backbone.

The positions of the Ara*f* residues in PMIa were determined by subjecting it to oxalic acid hydrolysis. After removal of the Ara*f* residues by this method, the amount of (1→3,6)-linked Gal*p* residues decreased while the amount of (1→6)-linked Gal*p* residues increased. This indicated that Ara*f* was linked to *O*-3 of the (1→6)-linked Gal*p* side chains in the original polysaccharide [2].

In the $^{13}$C-NMR spectrum of PMIa, the signals from terminally linked Ara*f* were dominating. Chemical shift assignments are given in Table 2. The resonances originating from Gal*p* were less well defined due to the occurrence of (1→3)-, (1→6)- and (1→3,6)-linked residues, but the signals from (1→3,6)-linked Gal*p* residues were of higher intensity. This confirmed the presence of a highly branched structure.

**Table 2.** Chemical shifts (ppm) of Ara*f* and Gal*p* residues from the $^{13}$C NMR spectra of PMIa and PMIa after weak acid hydrolysis recorded at 90°C (DMSO 39.6 ppm) [2]

| | Chemical shifts (ppm) | | | | | |
|---|---|---|---|---|---|---|
| | C-1 | C-2 | C-3 | C-4 | C-5 | C-6$^{d}$ |
| T-Ara | 110.1 | 82.2 | 77.4 | 84.7 | 62.1 | |
| (1→3,6)-Gal | 104.0 | 70.8 | 80.9 | 69.3 | 74.2 | 70.7 |
| (1→6)-Gal | 104.2 | 71.6 | 73.6 | 69.5 | 74.6 | 70.7 |
| (1→3)-Gal | 104.5 | 71.2 | 83.0 | * | 76.0 | 61.9 |

* co-occurred with the C-4 signals from (1→6)- and (1→3,6)-linked Gal*p* in both spectra.

$^{d}$ confirmed by DEPT.

After weak acid hydrolysis, a simpler spectrum was obtained. Resonances from (1→6)-linked Gal*p* residues were dominating, and the characteristic signals from Ara*f* were lacking.

In the $^{1}H$ NMR spectrum of PMIa, the anomeric signals of α-L-Ara and β-D-Gal occurred at 5.3 and 4.5 ppm, respectively. The latter had the characteristic coupling ($J_{1,2b}$ = 7.3 Hz) of a β-anomeric configuration [2].

## *Pectin*

PMII was the major IEC fraction isolated from the 50°C leaf water extract. It constituted about 37% of the carbohydrates in the crude extract and was composed of 9% arabinose, 8% galactose, 4% rhamnose, 71% galacturonic acid and 7% glucose. The molecular weight was estimated to 46-48 kDa [4]. Due to its monosaccharide composition, PMII is a pectin.

Pectins are composed of a galacturonan backbone with some insertions of rhamnose residues mainly located in regions of the molecule that contain side chains linked to some of the rhamnose residues. The side chains are often composed of arabinose and galactose. The regions containing side chains are referred to as the "hairy regions" [18] or "rhamnogalacturonan I" (RG-I) [19]. The homogalacturonan regions of the molecule are often called the "smooth regions".

**Table 3.** Chemical shift assignments for the $^{13}C$ NMR spectrum of the pectin PMII from the leaves of *Plantago major* L. The spectrum was recorded on Jeol DX270 at 80°C with a 1.2 sec pulse delay, 0.3 sec acquisition time and a 45° pulse angle, using 13.4K sweep width and 8K data points. Chemical shifts were set relative to DMSO (39.6 ppm) [3].

| | Chemical shifts (ppm) | | | | |
|---|---|---|---|---|---|
| | C-1 | C-2+3 | C-4 | C-5 | C-6 |
| GalA | 99.9 | 68.7-69.4 | 79.2 | 71.2 | n.r. |
| GalA-OMe | 100.6 | 68.7-69.4 | 79.2 | 71.2 | 171.1 |

n.r. = not resolved

In the $^{13}C$ NMR spectrum of PMII, only the signals from GalA and methyl esterified GalA (GalA-OMe) could be detected [3]. Assignments are given in Table 3 and were made by comparing with spectra of pectins from other sources [20]. The C-4 signal appeared at 79.2 ppm, showing that the

GalA residues in PMII were 1,4-linked. The presence of (1→4)-linked GalA residues was confirmed by reduction, methylation and GC-MS analysis [4].

The $^{13}C$ NMR spectrum also showed that this was a highly esterified pectin. There were two anomeric signals at about 100 ppm, one from esterified galacturonic acid and one from unesterified residues appearing at slightly higher field. The degree of esterification was 67%. This was determined by other methods; by $^{1}H$ NMR and by hydrolysis off methanol and quantification by GC [4].

Oligosaccharides consisting of (1→4)-linked GalA residues with estimated d.p. 4-10 were prepared by partial hydrolysis of PMII using 0.5M TFA [5]. The oligosaccharides were isolated by HPAEC-PAD. Four fractions containing oligosaccharides with both (1→4)-linked GalA and (1→2)-linked Rha were also isolated [5]. The oligosaccharides originate from the backbone of the pectin.

For isolation of the ramified regions, PMII was digested with pectinase after pectinesterase treatment. From the enzyme resistant part of PMII, two fractions, PVa and PVb corresponding to two different ramified regions were isolated by SEC using a Bio Gel P10 column [4]. Monosaccharide- and linkage analysis revealed that the fraction that was eluted in the void volume of the column, PVa, contained more neutral side chains linked to *O*-4 of the (1→2)-linked Rha residues in the backbone than the other fraction, PVb.

**Table 4.** The anti-complementary activity of PMII and PMII derivatives [4].

| | % anti-complementary activity 333 μg/ml |
|---|---|
| PMII | 73.3 ± 3.9 |
| PVa | 75.8 ± 1.0 |
| PVb | 61.3 ± 4.3 |
| Precipitate | 25.7 ± 4.3 |
| 0.5M TFA 5h | 24.0 ± 2.9 |

The side chains in PVa were composed mainly of terminally and 1,4-linked Gal*p* residues, (1→3)-linked Gal*p* residues with Ara*f* linked to *O*-6 and of (1→6)-linked residues with Ara*f* linked to *O*-3. The Ara*f* residues were terminally linked, (1→2), (1→3)- and (1→5)-linked [4]. The individual side chains were not isolated.

Structure-activity studies showed that PVa was the part of PMII that had the highest anti-complementary activity (Table 4). During enzyme digestion, a precipitate that had very low anti-complementary activity was formed. It contained mostly (1→4)-linked GalA residues. The activity of PMII decreased after hydrolysis with 0.5M TFA for 5h. Thus, the unbranched regions of PMII had low anti-complementary activity while the highly branched region PVa had high activity [4].

## 3. Heteroxylans from the seeds

The seeds were extracted with water at 50°C, and a crude extract with high viscosity was obtained. The extract was heated on a boiling water bath to reduce viscosity and fractionated by IEC and SEC [6]. The neutral IEC fraction, N contained starch due to positive reaction with iodine and a high content of (1→4)-linked Glc*p* residues.

Four acidic fractions, A, B, C and D were isolated using a DEAE-Sepharose fast flow column (Pharmacia) with a NaCl gradient (0-1M). IEC fraction A could be separated by SEC using a Sephacryl S-400 HR column (Pharmacia) eluted with 15 mM NaCl giving A1, A2 and A3.

**Table 5.** Monosaccharide composition (mol %) of polysaccharide fractions from *Plantago major* L. seeds. Reprinted from [6] with permission from Elsevier Science.

| | Fractions | | | | | |
|---|---|---|---|---|---|---|
| | A1 | A2 | A3 | B | C | D |
| Ara*f* | 27.6 | 15.0 | 17.8 | 12.6 | 15.6 | 15.9 |
| Xyl*p* | 37.5 | 21.7 | 23.5 | 39.0 | 53.3 | 52.8 |
| Rha*p* | trace | 3.7 | 3.5 | 3.4 | 2.1 | 2.9 |
| Gal*p* | 6.8 | 23.7 | 13.8 | 6.0 | 1.6 | 1.1 |
| Glc*p* | 4.0 | 4.2 | 26.4 | 3.5 | 0.6 | - |
| Man*p* | 1.6 | trace | 2.0 | - | 1.6 | - |
| Gal*p*A | 17.3 | 31.7 | 10.3 | 20.1 | 13.8 | 13.2 |
| Glc*p*A | 5.2 | trace | 2.7 | 15.6 | 13.1 | 14.1 |

Fraction A had the highest anti-complementary activity of the acidic fractions, fraction D had very low activity. A1 had higher activity than both A2 and A3 [6].

All the acidic fractions contained the same polysaccharide type, only differing in the relative amounts of various monosaccharides (Table 5).

The galacturonic acid residues were (1→4)-linked and may, together with galactose and rhamnose originate from a pectin present in the fractions.

The glucuronic acid residues were terminally linked while the arabinose residues were both terminally and (1→3)-linked in all fractions. The xylose residues were both terminally, (1→3)- and (1→4)-linked, and high amounts of the (1→4)-linked residues were also linked either at *O*-2 or at *O*-3. The fractions were therefore assumed to contain xylans that had both (1→3)- and (1→4)-linked Xyl*p* residues in the backbone and with a relatively high degree of branching of the (1→4)-linked residues [6].

$^{13}C$ NMR spectra of IEC fraction C confirmed the presence of both (1→3)- and (1→4)-linked Xyl*p* residues in the backbone. Some chemical shift assignments are given in Table 6. The high intensity signals from non-reducing ends of Xyl*p* indicated a high degree of branching. This was confirmed by the multiple signals appearing in the C-1 and C-5 regions of Xyl*p* [6].

**Table 6.** Chemical shift assignments for the $^{13}C$ NMR spectra of heteroxylan fraction C from the seeds of *Plantago major* L. [6].

| | Chemical shifts (ppm) | | | | |
|---|---|---|---|---|---|
| | C-1 | C-2 | C-3 | C-4 | C-5[a] |
| T-β Xyl*p* | 104.1 | 73.5 | 76.4 | 70.0 | 65.9 |
| (1→3)-β Xyl*p* | 104.1 | 74.0 | 84.3 | 68.4 | 65.8 |
| (1→4)-β Xyl*p* | 102.6 | 73.5 | 74.1 | 77.1 | 63.6 |

[a] confirmed by APT

Oligosaccharides were prepared from IEC fraction C by partial hydrolysis using 0.1M TFA. Fractions were isolated by HPAEC and analysed by total hydrolysis, methylation, GC-MS, ES-MS and ES-MS/MS [7].

(1→4)-linked Xyl*p* di- and trisaccharides were isolated, and (1→3)-linked Xyl*p* oligosaccharides with d.p. 2 and 6-11 were identified from the isolated fractions. This showed that the backbone of the molecule consisted of blocks of (1→4)-linked Xyl*p* and (1→3)-linked Xyl*p*, respectively. The largest (1→4)-linked block that was isolated contained two residues, and the largest (1→3)-linked block contained ten residues. According to the ratio of

(1→3)- and (1→4)-linked residues in the original polysaccharide, blocks containing more than two (1→4)-linked Xyl*p* residues probably exist [7].

A (1→2)-linked Xyl*p* disaccharide and a branched (1→4)-linked tetrasaccharide with a single Xyl*p* residue at the reducing end were also isolated, showing that single Xyl*p* residues were linked to *O*-2 of the (1→4)-linked residues in the backbone [7].

After removal of Ara*f* residues by weak hydrolysis with oxalic acid, the amount of (1→4)-linked Xyl*p* residues increased while the amount of (1→2,4)- and (1→3,4)-linked residues decreased. This indicate that Ara*f* residues may be linked directly to Xyl*p* in the backbone. The amount of (1→3)-linked Xyl*p* residues also decreased while an increase in non-reducing terminals of Xyl*p* was observed indicating the persence of Ara*f*-(1→3)-Xyl*p* side chains. Finally, a decrease in Glc*p*A content was observed, indicating that Glc*p*A was substituted to *O*-3 of Ara*f* in side chains [6]. The presence of Glc*p*A-(1→3)-Ara*f* side chains was confirmed by oligosaccharide analysis by total hydrolysis, methylation and GC-MS and ES-MS [7].

## 4. Concluding remarks

The polysaccharides isolated from *P. major* belong to polysaccharide classes that are normally found in plants. All higher plants contain pectin in the primary cell wall and in the middle lamella. Pectins are polysaccharides that are composed of smooth and hairy regions. The smooth regions in PMII had low anti-complementary activity while the ramified region PVa had high activity. PVa corresponds to the hairy regions or RG-I in pectins.

Arabinogalactans may be present as side chains in pectins, associated with the plasma membrane or in plant exudates. PMIa was classified as an arabinogalactan type II which is a polysaccharide type that is widely distributed in the plant kingdom [21].

Heteroxylan polysaccharides are found in other mucilage forming seeds, but having different fine structures. *E.g.* the mucilage from *Plantago ovata* contains a heteroxylan with both (1→3)- and (1→4)-linked residues in the backbone having other types of side chains than those found in *P. major* [22]. The heteroxylans from *P. asiatica* and *Linum usitatissimum* on the other hand contain only (1→4)-linked Xyl*p* residues in their backbones [23, 24]. Previously it has been shown that the acidic xylan from *P. asiatica* has high anti-complementary activity [16].

At present time it is not known whether all RG-I, arabinogalactans type II and heteroxylan polysaccharides possess the same type of biological activity. The great challenge is to find the minimum structures of these polysaccharide types that are required for expressing anti-complementary activity.

## References


[1] Samuelsen, A.B., Paulsen, B.S., Wold, J.K., Otsuka, H., Yamada, H. and Espevik, T. (1995). *Phytother. Res.* 9, 211-218.

[2] Samuelsen, A.B., Paulsen, B.S., Wold, J.K., Knutsen S.H. and Yamada, H. (1998). *Carbohydr. Polym.* 35, 145-153.

[3] Samuelsen, A.B., Knutsen, S.H., Cohen, E.H., Paulsen, B.S. and Wold, J.K. (1995). Poster B-33 at *The 8th Eurpoean carbohydrate symposium*, Seville, Spain.

[4] Samuelsen, A.B., Paulsen, B.S., Wold, J.K., Otsuka, H., Kiyohara, H., Yamada, H. and Knutsen, S.H. (1996). *Carbohydr. Polym.* 30, 37-44.

[5] Samuelsen, A.B., Cohen, E.H., Paulsen, B.S. and Wold, J.K. (1996). In *Pectins and Pectinases*, Visser and Voragen (Eds.). Elsevier Science, Amsterdam, pp. 619-622.

[6] Samuelsen, A.B., Lund, I., Djahromi, J., Paulsen B.S., Wold, J.K. and Knutsen, S.H. (1999). *Carbohydr. Polym.* 38, 133-143.

[7] Samuelsen, A.B., Cohen, E.H., Paulsen, B.S., Brüll, L.P. and Thomas-Oates, J. E. (1999).*Carbohydr. Res.* Accepted.

[8] Roca-Garcia, H. (1972). *Arnoldia* 30, 23-24.

[9] Nielsen, H. (1969). *Lægeplanter og trolddomsurter*. Kehler, S. (Ed.). Politikens Forlag A/S, København, pp. 321-324.

[10] Lim-Sylinaco, C.Y. and Shier, W.T. (1985). *J. Toxicol. toxin reviews* 4, 71-105.

[11] Jain, S.K. (1991). *Dictionary of Indian folk medicine and ethnobotany.* Deep Publications, New Delhi, pp. 145.

[12] Høeg, O.A. (1974). *Planter og tradisjon.* Universitetsforlaget, Oslo, pp. 507-511.

[13] Brøndegaard, V.J. (1987). *Folk og flora.* Vol. 4, Rosenkilde og Bagger, pp. 68-77.

[14] Hussey, J.S. (1974). *Econ. Bot.* 28, 311.

[15] Guillén, M.E.N., Emim, J.A.S., Souccar, C. and Lapa, A.J. (1997). *Int. J. Pharmacognosy* 35, 99-104.

[16] Yamada, H., Nagai, T., Cyong, J.C., Otsuka, Y., Tomoda, M., Shimizu, N. and Shimada, K. (1985). *Carbohydr. Res.* 144, 101-111.

[17] Aspinall, G.O. (1973). In *Biogenesis of plant cell wall polysaccharides.* Loewus, F. (Ed.). Academic Press, New York.

[18] De Vries, J.A., Rombouts, F.M., Voragen, A.G.J. and Pilnik, W. (1982). *Carbohydr. Polym.* 2, 23-33.

[19] Lau, J.M., McNeil, M., Darvill, A.G. and Albersheim, P. (1985). *Carbohydr. Res.* 137, 111-125.

[20] Westerlund, E., Åman, P., Andersson, R.E. and Andersson, R. (1991). *Carbohydr. Polym.* 14, 179-187.

[21] Clarke, A.E., Anderson, R.L. and Stone, B.A. (1979). *Phytochemistry* 8, 521-540.

[22] Sandhu, J.S., Hudson, G.J. and Kennedy, J.F. (1981). *Carbohydr. Res.* 93, 247-259.

[23] Tomoda, M., Yokoi, M. and Ishikawa, K. (1981). *Chem. Pharm. Bull.* 29, 2877-2884.

[24] Muralikrishna, G., Salimath, P.V. and Tharanathan, R.N. (1987). *Carbohydr. Res.* 161, 265-271.

# Chapter 5

# Pharmacological Activities of Sulfated Carbohydrate Polymers

G. FRANZ, D. PAUPER and S. ALBAN
*Institute of Pharmacy, University of Regensburg, Universitätsstraße 31, 93040 Regensburg, Germany*

Key words: Heparinoids, Glucan sulfates, Anticoagulant Activity, Angiogenesis inhibitors, Galactan sulfates, CAM Assay

Abstract: Among the large group of naturally occuring polysaccharide-sulfates, the glycosaminoglycans (GAG) such as heparin, heparansulfate, chondroitin-sulfate, dermatansulfate and keratansulfate are biologically important polymers. The best known of these compounds is heparin, where the position of the sulfate groups is essential for any anticoagulant effect. In awareness of various contaminants in natural products of animal origin, there is increasing fear that many products from bovine or lifestock origin many be contaminated. This has stimulated the development of synthetic heparin-analogs (heparinoids) such as the heparin pentasaccharide and others. Another source for sulfated polysaccharides are the sulfated biopolymers of marine origin such as the carrageenans, fucoidans, fucans and other algal cell wall polysaccharides, which in part have been examined for anticoagulant and antithrombotic effects. Suflated polysaccharides from bacterial origin are of special drug innovation intererst, since large amounts of the starting polymer can be obtained by biotechnological methods. Another group of sulfated polysaccharides consists of the semisynthetic polysaccharide sulfates such as dextransulfate, pentosan-polysulfate, chitosansulfate and finally β-1.3 glucansulfates.

The group of linear or slightly branched glucans of bacterial origin was chosen for studies about the mechanisms of polysaccharide-sulfate protein interactions in the series of physiological events in the coagulation cascade. With sulfated β-1.3 glucans, differing in DS, sulfate position and chain length, it was tried to establish an ideal anticoagulant polymer which could serve as valuable lead-structure for the future development of an optimal carbohydrate based antithrombotic drug.



*B.S. Paulsen (ed.), Bioactive Carbohydrate Polymers,* 47–58.

Besides antithrombotic and antiviral effects, sulfated polysaccharides are known to inhibit the complex process of angiogenesis, i.e. the growth and development of new blood vessels, which occurs in many tissues and is a prerequisite of tumor progression. Tumor growth and metastasis is angiogenesis dependent when an increasing number of new capillaries converge upon a tumor. Inhibition of angiogenesis, without cytotoxic side effects, is one of the new approaches of tumor treatment. The basic fibroblast growth factor (bFGF) has been shown to be one of the most potent angiogenic molecules. Since it is known that heparin and heparin-like structures have a high affinity towards bFGF and thereby act as a regulator of its function, the search for other heparin-like polymers was initiated. Heparin itself has the disadvantage of uncontrolled bleeding risk when applied in a systemic treatment.

A series of other sulfated polysaccharides differing in their basic structures (primary structure, type of glycosidic linkage, degree of sulfation, chain length) was examined in the CAM-assay.

Genuine carrageenans and modified carrageenans were tested for their ability to inhibit the formation of new blood capillaries with good success, however, they had considerable side effects. Better results were obtained with a series of modified galactans. It could be shown that β-1.4-galactansulfates of relatively low DP with a distinct DS have very interesting antiangiogenic activity. These non-cytotoxic galactan structures are further developed as angiogenesis inhibitor lead-structures.

# 1 Introduction

During the past decade many significant developments in the utilization of carbohydrate polymers as drugs have been put forward. This, however, is not surprising since it is known that many of these biopolymers play an essential role in key biological processes. Some of the new fields of polysaccharide applications have been approved by the scientific world, others are still a matter of controversies and hence are not being accepted for clinical approval [1]. An essential prerequisite when distinct physiological effects are attributed to a natural compound, is the knowledge of the exact structural parameters. This means in the case of polysaccharides, the type of the glycosidic linkage between the individual sugar components, the degree of branching, the molecular dimensions and the possible conformation of the macromolecule [2]. Furthermore it is of high importance to learn about the possible substituents which can either influence the polarity or the ionic charge of the polymer in question.

As in the case of the widespread glycosaminoglycans (GAG), it was shown that several types of functional groups such as sulfate esters, carboxylate, acetamide, primary and secondary hydroxyl groups can be present.

The biological function of the biopolymer is related to precise distribution of these functional groups on sometimes well defined oligosaccharide sequences, as it has been shown for heparin when binding to antithrombin [3]. Other functions of this highly anionic biopolymer could not be asigned to precise structures like the antiviral activity, binding to growth factors such as bFGF or the smooth muscle cell growth inhibition.

Besides heparin there is a great number of other sulfated biopolymers in the animal- and plant kingdom to which only in part distinct physiological functions could be attributed. Some of these polymers have been examined in the past, but many still await a physiological/pharmacological screening in order to demonstrate possible lines of medicinal application. Chemical modification or partial synthesis in order to obtain new structures, i.e. sulfated polysaccharides is an additional chance to obtain new pharmacological agents for a possible therapeutic use [4].

## 2. Recent development in anticoagulant/antithrombotic agents

The newer developments in antithrombotic therapeutics are rather significant. Many advanced techniques to develop antithrombotic drugs are used at the present time. Advances in biotechnology and separation techniques have also contributed to the development of new therapeutic anticoagulant systems [5].

The classical agent for the prevention and treatment of thrombosis is unfractionated heparin. Usually, beef lung and porcine mucosal derived products are available, which only differ slightly in their composition. All unfractionated heparins are heterogeneous in nature. The molecular weight varies depending on their origin and mode of extraction. The beef lung heparins are of higher molecular weight compared to the porcine mucosal heparin. However, the anionic nature of both compounds is comparable. Chemical modifications of the genuine polydisperse heparin such as desulfation or deamination resulted in products devoid of anticoagulant activity but still with selective interactions on enzymes, cellular receptors or virus proteins [5]. These compounds are currently being tested for virucidal activities and treatment of proliferation disorders [6].

## 3. Low molecular weight heparin (LMWH)

The development of short chain better defined heparin, i.e. LMWH, has opened new dimensions in thrombosis prevention and treatment. Mainly their relative effects on platelets are reduced compared to heparin. The observation that only LMWH is absorbed by the body after subcutaneous administration has been a real breakthrough [5].

The consequence is that LMWH is much better bioavailable than unfractionated heparin. Furthermore, the LMWH's exhibit a longer biologic half life in contrast to heparin, with the consequence that LMWH-preparations can be administered as a single day injection as a prophylactic agent [7].

Degradation under controlled conditions of genuine heparin is possible by different chemical and enzymatic procedures. However, recent studies have shown that the individual LMWH's on the market exhibit chemical and pharmacological differences.

## 4. Heparinoids: heparin analogues

As a result of the heparins associated multiplicity of biological activities, further the possible contamination with BSE or porcine disease, substantial research effort has been dedicated to the discovery of heparin analogues, which might be obtained from animal, plant, microbial or synthetic sources. Since it is no longer believed that sulfated polysaccharides must exhibit interactions with antithrombin III to have effective antithrombotic properties, several agents differing considerably from heparin in their respective chemical structure have been developed [4].

Many efforts were based on dermatan sulfate with the result of some products on the market which differ considerably in the pharmacological aspects compared to heparin. It is interesting to note that despite of similar composition, various companies are making these agents available for different clinical applications.

In the field of newly synthesized anticoagulant compounds, the synthetic design of the antithrombin III binding pentasaccharide was a real breakthrough [8]. However, the multistep-synthesis is still rather uneconomic and some of the features specific for heparin are missing. Actually some pentasaccharide analogues are under clinical investigation.

## 5. Plant derived sulfated polysaccharides

Not only GAG of animal origin but also algal polysaccharides are a potential source of heparin like substances. Sulfur is the fourth element in marine water, and the sulfate ion is the most stable combination of sulfur in seawater. Consequently it is not surprising to find many sulfated compounds in marine organisms [9].

Carrageenans are one of the longest known samples of algal sulfated galactans with some activities as anticoagulants. But the structure of the anionic polymers in questions show a broad variability with the consequence of nonconstant physiological activities. In general, the λ-carrageenan with the highest degree of sulfation is the most potent anticoagulant followed by κ-carrageenan and τ-carrageenan [10].

Fucoidan is typical for brow algae, composed of highly sulfated glucoronofucoglycans and homofucans. Due to inconsistent structures and molecular dimensions, marked differences are obvious in their physiological activities. Furthermore, fucan sulfates have a mode of action different from that of animal GAG's such as heparin [11]. The most important difference is the inhibition of fibrin polymerization by thrombin. Further sulfated polysaccharides from red and brown algae have been analysed under a chemical viewpoint and pharmacological effectiveness. But none of these marine polymers has gained medical importance.

## 6. Semisynthetic-Polysaccharide sulfates

Since the natural occuring sulfated biopolymers were not fully satisfying as heparin substitutes, it was tried to start with well defined non sulfated polymers and to introduce stepwise, under controlled conditions, the respective sulfate groups thereby controlling the degree of sulfation (DS) and the position of the sulfate groups on the different C-atoms of the monomer units.

A large series of neutral and acidic, linear and branched, low and high molecular weight genuine carbohydrate polymers have been sulfated by means of the classical methods. A pretreatment with DMF (N.N-dimethylformamide) seems to improve the reaction, since DMF associates with the free hydroxylgroups of the polymer and makes them more accessible to the sulfating reagent, i.e. the $SO_3$/pyridine complex. The concentration of the $SO_3$/pyridine mixture, the reaction time and the reaction temperature are important parameters in order to influence the DS of the final compound [12].

Control of the non-destructive reaction conditions was carried out by GPC-measurements. Determination of the substitution pattern can be done by the classical methylation procedure and analysis of the methylated/acetylated compounds by GC/MS. The sulfate groups and the glycosidic linkages can be cleaved by a treatment with TFA [13] [14].

By this way a series of different β-1.3 glucans of bacterial and algal origin were derivatised and analysed, further the branched glucans pullulan and the highly branched xanthan were included in this procedure [15].

The newly synthesized polysaccharide sulfates were examined in the classical coagulation assays in order to examine their specific interaction in the complex of the blood coagulation cascade.

The thrombin time (TT) determines the inhibition of the thrombin mediated fibrin-formation, i.e. the anti-IIa-activity. The Heptest® is determining the anti-Xa-activity and the aPTT reflects the activity of the intrinsic pathway.

A minimum DS for most of the sulfated polymers seems to be essential to be active in all three assays. It varies in dependence on the type of the glycosidic linkage, which certainly has some influence due to the specific conformations of β-1.3 or β-1.4 glucans. The increased activity as shown for the sulfated α-1.4/1.6 glucans (pullulan) compared to a pure β-1.3 glucan (curdlan) is certainly correlated with the higher flexibility of the macromolecule, which is essential for any interaction with the different coagulation factors. The influence of the DS is clearly documented in a bell-shaped curve with a maximum inactivity at about 1.5. The optimum curve was again obvious for the molecular weight influence determined for polymer-populations with a constant DS [12].

A maximum effect was obvious in a MW range of 40-60 kd. Some differences in the three test systems, however, were obvious. The anticoagulation mechanism which is essential for a prolonging of the clotting time in the Heptest® is favoured by shorter chains compared to the TT, caused by an interference with Thrombin mediated fibrin formation, which requires a certain minimum chain length [12].

As a consequence of the findings observed in vitro, it was of interest to demonstrate in vivo antithrombotic activity of these newly synthesized heparinoids. However, animal models are of limited value when interpreting the results for human validity, but they provide for patients indications about the overall effects in living systems. Two test models were employed: the FEIBA® induced stasis thrombosis rabbit model and a clamping induced occlusion rat model. In the first model, after the application of the test substance, a hypercoagulable state is induced by the injection of FEIBA® followed by the ligation of the jugluar vein. After a certain time of stasis, the ligated vessel segments are dissected and the score of the formed clot, representing the antithrombotic activity is determined. In this model, thrombosis is induced via endothelial injury [16].

In the second rat model, the repeated clamping of the jugluar vein results in stasis as well as in injury of the vessel wall with consequent adhesion and activation of platelets. In this assay, the antithrombotic activity is reflected by the number of clampings required for the interruption of the blood flow determined by doppler ultrasound [16].

In the rabbit model, the tested heparinoids produced concentration dependent antithrombotic actions with ED values considerably higher than that of heparin. However, compared to the weak effects in the rabbit model in the rat-assay the sulfated glucans were shown to produce heparin like antithrombotic activities. At a dosage of 250µg/kg both agents showed similar effects. Consequently, the new glucan based heparinoids are shown to be potent antithrombotic compounds including not only the coagulation systems but also the platelets and the vessel wall.

Conclusion: The results clearly demonstrate that these new, well defined glucan sulfates are potent anticoagulant compounds with a high antithrombotic potential. For additional tests of these agents, different assays and models should be considered to determine their differential antithrombotic actions.

## 7. Sulfated carbohydrate polymers as angiogenesis inhibitors

Angiogenesis, which is defined as the growth in initiation of new blood vessels, occurs in most tissues and can be induced by cytokines. Angiogenesis is fundamental to reproduction development and repair mechanisms. Under these conditions, angiogenesis is highly regulated and of short duration. In some pathological situations this regulation is deranged with the consequence that the disease is driven by peristent neovascularisation [17].

Although it has been known for many decades that tumors contain an abnormally dense blood vessel network, it was only about 25 years ago when it was realized that tumors induce their own blood supply, which depends upon the overall process of angiogenesis. According to Folkman and coworkers, tumor growth is parallel to vascularization and after this process, exponential in growth. Angiogenesis provides tumor cells with a ready access to the blood circulation, which assists metastasis [18].

However, angiogenesis is not limited to malignant tumors, it is also existent in non-neoplastic diseases and some normal process. Non-neoplastic diseases comprize rheumatoid arthritis, hemangiomes, angiofibromas, psoriasis, atherosclerosis and finally eye diseases such as diabetic retinopathy and neurovascular glaucoma. Angiogenesis occurs in healthy adults and its evident during wound healing, ovulation, menstruation and pregnancy [19].

Angiogenesis is a complex process in which existing mature vessels generate sprouts that develop into complete new vessels.

During tumor angiogenesis, vascular cells grow at abnormally rapid rates. Angiogenesis involves three major events:

The basement membrane surrounding a mature capillary dissolves; onset of growth out of the capillary

A hollow sprout generates from the bud and continues toward the angiogenic stimulus

The sprout joins its end with another sprout to form new capillary vessels.

The overall process involves activation, adhesion, migration, proliferation and transmigration of endothelial cells across cell matrices to or from new capillaries.

Endothelial cells play a major role in the modelling of blood vessels. In mature organisms and under normal physiological conditions, endothelial-cell turnover and angiogenesis rates are extremely slow. This process can be activated in situations such as wound healing and ovulation. The interplay of growth factors, cell adhesion molecules and specific transduction pathways is critical under physiological or in pathological angiogenesis conditions [19].

## 8. Angiogenic mediators

Numerous angiogenic mediators have been discovered. They vary in the way they induce angiogenesis and may act independently or in contact with another. They can stimulate growth of endothelial cells, attract macrophages to the site which then secrete angiogenic factors.

Basics fibroblast growth factor (bFGF) is one of the potent angiogenic factors besides the vascular endothelium growth factors (VEGF) and others [20].

The specific inactivation of these factors may be one important step for a specific blocking of this process. However, there are still more mechanisms which can be influenced in order to specifically interrupt the overall process of angiogenesis.

Besides the angiogenic stimulus, the infraction rate of endothelial cells, the proteolysis of the basal membrane, the lumen formation and receptor binding of angiogenic factors such as bFGF.

The angiogenesis inhibitors which have been successfully tested in different models show a broad spectrum of chemically different structures such as steroids, polypeptides, antibiotics and anionic polysaccharides such as heparin, heparan sulfate, pentosan polysulfate and sucralfat [21].

These anionic carbohydrates interact by binding the relevant growth factors. It is well documented that many growth factors possess heparin binding sites. For pentosan polysulfate it was shown to bind FGF, further to interact in vivo with angiogenesis and thereby reducing significantly tumor growth.

However, the drawback with heparin and all the other sulfated polysaccharides tested so far, is the pronounced anticoagulating activity, which is responsible for uncontrolled bleeding, when administered as antiangiogenic substance. It was therefore tried to establish anionic polymers with reduced anticoagulating activity but still pronounced antiangiogenic capacity. One further problem, when dealing with sulfated polysaccharides is the fact that all these biopolymers after isolation and purification are polydisperse with the consequence that one is dealing not with one single macromolecule but with a range of different chain length representatives, which only have in common an overall comparable structure.

Since it is known that the anticoagulating activity of for example heparin, largely depends upon the chain length, a similar phenomenon was expected from other sulfated polysaccharides [22]. However, one advantage of these biopolymers is the very low toxicity and tolerance even at high dose applications. This excellent tolerance of almost all of the tested sulfated polysaccharides was obvious in the basic test system, which served as most sensitive screening model: i.e. the CAM-assay. This model is very sensitive for all kinds of cytotoxic and embryotoxic effects. With most of the applied substances no unspecific cytotoxic effects were shown [23].

## 9. CAM-Assay

In this test system fertilized hen eggs are incubated for three days and then opened with the necessary precautions, and the shell is separated from the residual living egg. The embryos are placed in a sterile in vitro system or maintained under semi sterile conditions for 2 days at 37°C and 70% rel. humidity at which point the treatment is started to assay angiogenesis inhibition. The test compounds are inbedded intro agarose pellets for a sustained release. At this point the CAM = Chorioallantoismembrane is developed and is forming vessels under normal conditions. The average diameter of the membrane is about 2 cm. The agarose pellets are in direct contact with the newly formed membrane without any physiological interference, which is shown by the respective control experiments. As positive control, usually suramine, a potent but highly toxic angiogenesis inhibitor is utilized.

## 10. Quantification of the angiogenesis inhibiting effect

Normally after 48h the antiangiogenic response is observed with the aid of a stereo microscope. The positive or negative effect is easy to detect and to evaluate. However, when assaying a dose dependence or evaluating relatively weak effects, the quantification of angiogenesis inhibition is difficult to document. A minimum of 10 eggs is included in parallel in each assay for comparison of the individual drug concentrations.

The chicken embryos are coded and examined randomly. A certain degree of quantification is made possible after establishing a score system in which it is possible to distinguish between zero, weak and strong effects. The score values were as follows:

0: no effect

1: weak to medium effect; i.e., small capillary free areas

2: strong effect: the capillary free area is at least double the size as the drug containing pellet.

Thus, a score value of for example 0,4, may represent that 40% of the eggs showed the value 1 and 20% of the eggs obtained the value 2.

If one compound was subjected different times to the CAM assay, the resulting difference in the score values obtained was not greater than 0,1-0,2.

## 11. Testing of Galactans

In the following, a series of structurally different sulfated galactans were subjected to the CAM assay. In order to compare the effect of the more substituted, genuine polysaccharide structures, these polymers were tested as well.

It could be shown that neutral unsubstituted 1.4 galactans had almost no effect. However, when these biopolymers contained a certain degree of uronic acids, as it is the case of a galactan IS and Gal II, some activity was obvious. Polygalacturonic acid on the other hand was inactive, demonstrating that not only the charged anionic group is responsible for the physiological effect but in addition other structural parameters.

In a first series of experiments, the influence of the degree of sulfatation upon angiogenesis was tested. For this purpose, a linear β-1.4 galactan was selected, which, according to the method of sulfation, resulted in fractions with increasing DS from 0.14 to 1.70. The antiangiogenic effect was obvious starting from a DS of 0.50 [24].

When the possible influence of the molecular dimension of these sulfated polymers was examined, first with the high sulfated β-1.4 galactan and further with a similar sulfated lichenan and finally with carrageenan, no influence of the MW of these compounds was obvious. It could be concluded that at least in the range of 20 up to 1000 sugar monomers, the antiangiogenic effect seems to be independent of an existing chain length.

A next point of interest was the influence of branching in the existing galactan chains, companing the strictly linear LUPS β-1.4 galactan with the highly branched Asa PS and further the branched galactan, all with an almost similar DS. No distinct influence of the branching on the antiangiogenic effect was obvious [25].

In order to find out about the influence of specific sulfation patterns upon the antiangiogenesis effect, the natural differences in sulfation of κ, λ, τ, θ–carrageenans were examined. κ-carrageenan is almost exclusively sulfated in position 4, λ-Carrageenan in positions 4 and 2. Some influence can be seen between these different sulfation patterns. Most likely the more complete the anionic character around the sugar molecule, the better the antiangiogenic effect.

Additional experiments with selectively sulfated celluloses: Cellulose-2-sulfate, Cellulose-3-sulfate and Cellulose-2,3-sulfate confirmed the conclusion that all sulfate positions are possible in interacting with angiogenesis factors.

In order to examine the specificity of sulfate groups in comparison to other substituents i.e. galactan esters of a series of inorganic esters, nitrated, phosphorylated and fluorinated compounds were prepared and subjected to the CAM assay. The anionic effect of all compounds seems to have at least some influence whereby the sulfated galactan with a comparable degree of substitution seems to have the optimal effects [25].

## 12. **Conclusion:.**

Galactansulfates are potent nontoxic angiogenesis inhibitors, where dose dependence and structure/substitution related effects can be demonstrated. They do not interfere with blood coagulation as it is the case for other highly sulfated polymers. Since it has been demonstrated that a high MW is not essential for the antiangiogenic effect, relatively short chain galactan sulfates, which do not have the disadvantage of high viscosity and gel formation, are candidates for a further development of lead-structures. A next step will be the in vivo testing in animal tumor models in view of a tumor growth inhibiting effect.

## References

[ 1] Witczak, Z.J. (1995) Curr Med Chem 1: 392-405

[ 2] Franz, G., Alban, S., Kraus, J. (1995) Macromol Symp 99: 187-200

[ 3] Lindahl, U., Thunberg, L., Bäckström, G., Riesenfeld, J. (1984) J Biol Chem 259: 12368-12379

[ 4] Alban, S. (1996) In: Carbohydrates in Drug Design. Witczak, Z. J., Nieforth, K.A.,eds. Marcel Dekker Inc New York 209-276

[ 5] Fareed, J., Callas, D.D., Hoppensteadt, D., Jeske, W., Walenga, J.M. (1995) Exp. Opin. Invest. Drugs 4: 389-412

[ 6] Zacharski, L.R., Ornstein, D.L. (1998) Thromb Haemost 80: 10-23

[ 7] Kakkar, V.V. (1995). Thromb Haemost 74: 364-368

[ 8] Sinay, P., Jaquinet, J.C., Petitou, M., Duchaussoy, P., Lederman, J.,Choay, I., Torri, G. (1984) Carbohydr Res 132: C5- C9

[ 9] Kornprobst, J.M., Sallenave, C., Barnathan, G. (1998) Comp Biochem Physiol 119B: 1-51

[10] Güven, K.V., Özsoy, Y., Ulutin, O.N. (1991) Bot Mar 34: 429-432

[11] Nishino, T., Nagumo, T. (1992) Carbohydr Res 229: 355-362

[12] Alban, S. (1993) PhD Thesis, University of Regensburg, Germany

[13] Alban, S., Franz, G. (1994) Pure and Appl Chem 66: 2403-2406

[14] Alban, S., Franz, G. (1994) Seminars in Thrombosis and Hemostasis 20: 152-158

[15] Schauerte, A. (1999) PhD Thesis, University of Regensburg, Germany

[16] Alban, S., Jeske, W., Welzel, D., Franz, G., Fareed, J. (1995) Thromb Res 78: 20-210

[17] Folkman, J. (1993) C R Acad Sci Paris 316: 914-918

[18] Folkman, J., Klagsbrunn, M. (1987) Science 235: 442-447

[19] Paper, D.H. (1998) Planta Medica in print

[20] Hoffman, R., Paper, D.H. Ronaldson, J., Alban, S., Franz, G. (1995) J Cell Sci 108: 3591-3598

[21] Arbiser, J.L.(1997) Drugs of Today 33: 687-696

[22] Linhardt, R.J., Toshihiko, T.(1996) In: 'Carbohydrates in Drug Design'. Witczak, Z.J., Nieforth, K.A. eds. Mardel Dekker New York 277-339

[23] Paper, D.H., Vogl, H., Franz, G. (1995) Macrom Symp 99: 219-255

(24) Vogl, H., Paper, D.H., Franz, G. Carbohydr Polymers (in print)

[25] Vogl, H. (1996) PhD Thesis University of Regensburg, Germany

# Chapter 6

# Relationships between chemical characteristics and anticoagulant activity of low molecular weight fucans from marine algae.

F.CHAUBET (1)*, L.CHEVOLOT (2), J.JOZEFONVICZ (1), P. DURAND (2) and C. BOISSON-VIDAL (1)
*Unité de Recherches Marines n°2, CNRS – IFREMER, France.*

*(1) Laboratoire de Recherches sur les Macromolécules, CNRS, UMR 7540, Institut Galilée, Université Paris-Nord, Av. J.B. Clément, 93430 Villetaneuse (France).*
*(2) Laboratoire de Biochimie et Molécules Marines, IFREMER, B.P. 1105, Rue de l'Ile d'Yeu, 44311 Nantes cedex (France).*

* *to whom all correspondance should be addressed*

Key words: fucans, sulfate, anticoagulants

Abstract: Anticoagulant activity of fucans is closely linked to their chemical composition. The anticoagulant properties of various fractions extracted from different algae species were studied to assess the relationship between fucan structure and activity. For this purpose, cell walls were isolated from the brown algae and then submitted to sequential chemical extraction leading to high molecular weight fucans (HMWF). Different low molecular weight fucans (LMWF) were obtained by acidic degradation or radical process depolymerization of HMWF. Part of them were fractionated according to their molecular weight and charge density. We have demonstrated that the anticoagulant activity mainly correlated to the sulfate groups content and to the molecular weight distribution. The best results were obtained with fractions extracted from *Ascophyllum nodosum*. The anticoagulant and antithrombotic activities of each fraction were studied *in vitro, ex vivo* and *in vivo*. Affinity chromatography on stationary phase coupled with heparin binding plasma proteinase inhibitors was påerformed on the most active fractions. The high affinity fragments were purified and identified mostly sas α (1,2) and α (1,3)-linked units of 4-sulphuryl-fucose with sulfate or some branching at position 2 or 3.



*B.S. Paulsen (ed.), Bioactive Carbohydrate Polymers,* 59–84.

# 1. INTRODUCTION

Sulphated polysaccharides extracted from seaweeds represent a source of marine compounds with potential applications in medicine [1,2]. Mention of the use of seaweed products is found in Traditional Chinese Herbal Medicine as early as the sixteenth century [2,3]. They were mainly used in the treatment of sprains, rheumatism, cancer, bronchitis or emphysema and goitre [2]. Their anticoagulant properties were first described in extracts 50 years ago. Extracts from over 60 species covering brown, red and green seaweeds have been reported to exhibit anticoagulant capacity with the major active components identified as sulphated polysaccharides [3] although not all sulphated carbohydrates possess anticoagulant activity [4-6].

In 1936, Chargaff *et al.* reported for the first time on the anticoagulant properties of extracts from the red algae *Iridaea laminaroides* [7]. The anticoagulant material was described as a sulphated galactan. Further studies described similar properties in agar and carrageenan [8]. However, their use for clinical purpose appeared to be limited due to their high immunogenicity and gel formation tendency in relation respectively with their structural heterogeneities and very high molecular weight [1,2]. Concerning the green algae, in 1985 Deacon-Smith *et al.* have described the anticoagulant properties of crude aqueous extracts of *Codium fragile* [9]. The active fraction was a high molecular weight sulphated proteo-xylo-arabino-galactan [9,10]. More recently extracts *Codium latum* exhibiting very high anticoagulant activities were identified as sulphated polysaccharides mainly containing L-arabinose [11].

Sulphated polysaccharides from brown seaweeds have been the most extensively studied. They were first isolated by Kylin [12] from *Laminaria digitata* and *Fucus vesiculosus* and were named fucoidin when L-fucose (6-deoxy-L-galactose) was identified as an hydrolyzate [13,14]. 45 years later, Springer *et al.* described significant *in vitro* and *in vivo* anticoagulant activity of a fucan fraction named fucoidan isolated from *Fucus vesiculosus* [15]. From this time, many studies have been carried out on the preparation and the structure-activity relationships of fucan fractions from brown algae endowed with anticoagulant capacities.

# 2. HEPARIN AS A MODEL

In coagulation monotoring, the most used polysaccharide over the past 40 years is heparin. This glycosaminoglycan of animal origin was discovered by McLean in 1916 [16] and was used in clinical trials for the first time in 1937 [17]. Crude heparin is a linear polysaccharide formed by chains of O- and N-sulphated uronic acids and glucosamine (partially N-acetylated) [18] (Fig 1). Heparin exerts its main anticoagulant activity by potentiating the inhibitory

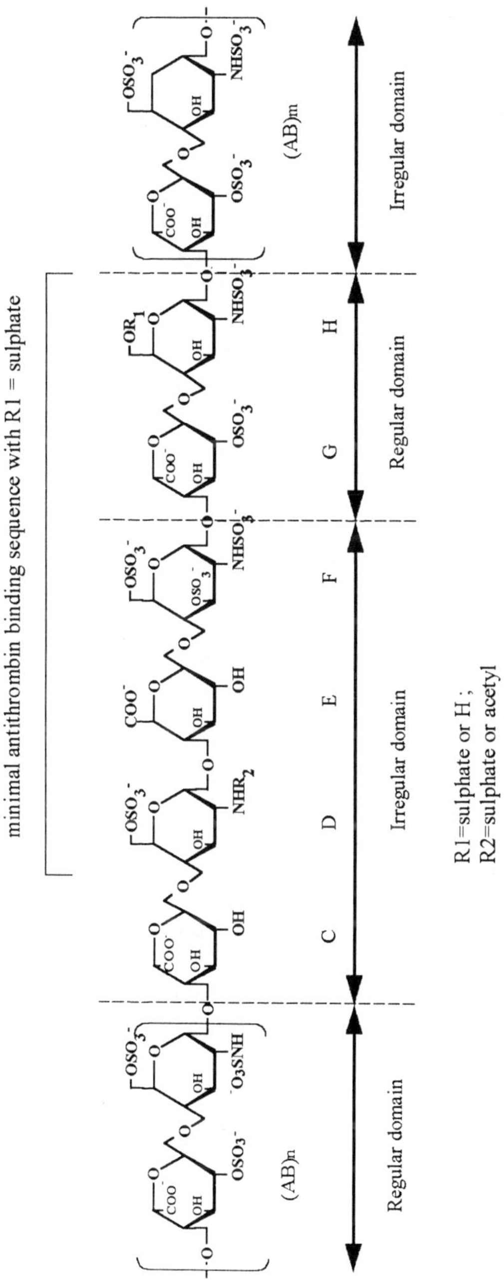


**Figure 1**
Schematic structure of heparin (from Casu *et al.* [4])

effect of the plasma serine protease inhibitors of the blood coagulation antithrombin (AT) and, to a much less extent, heparin cofactor II (HCII) [5,19-23]. The anticoagulant activity of heparin is due mainly to a specific pentasaccharidic sequence which binds with high affinity to AT, and thus inactivates enzymes involved in the coagulation process [4,21,24-27] (Fig 1). In recent years, heparin has been shown to be a heterogeneous polysaccharide with a high degree of polydispersity which diplays a wide variety of other biological effects unrelated to its anticoagulant activity. Over the past 15 years, fractionated and depolymerized forms of heparin (termed low molecular weight heparins) were developed [28,29]. These new compounds keep their high antithrombotic efficacy *in vitro* and in various venous thrombosis models without any anticoagulant effect at the concentrations used. Moreover, they are better adsorbed and have a longer half life [30]. However allergies and induced thrombocytopenia remains major drawbacks in the clinical use of heparin [31,32].

Numerous natural or hemisynthesized sulphated polysaccharides possess some of the biological activities of heparin [1]. Glycosaminoglycans such as hyaluronic acid, keratan sulphate, sulphated chondroitin, and sulphated polysaccharides of plant origin such as carrageenans and galactans have shown very low anticoagulant activities *in vitro* or *ex vivo* [3,10,33,34]. Some others sulphated polysaccharides showed promising results: 1-sulphated and 6-sulphated chondroitins [3], dextran sulphate and some of its derivatives [35-37]. Heparan sulphate, dermatan sulphate and pentosan polysulphate are among the most studied compounds [38-46]. A promising family of polysaccharides is fucans from marine origin. As by-products of alginate preparation in the food and cosmetic industries, fucans represent a cheap source of molecules of medical interest free of the risk of viral contamination associated with mammalian products and with less chance of side-effects such as bleeding and thrombocytopenia. Moreover, fucans (as does heparin) exhibit numerous other biological properties such as antiinflammatory [47,48], antiviral [49,50] and antiangiogenic [51]. They can modulate cell adhesion [52,53], growth factor release [54], smooth muscle cells proliferation [55], clinically relevant events such as tumor metastasis [1,54,56-58], and block sperm-egg binding in various species [59-61].

## 3. SOME STRUCTURAL FEATURES OF FUCANS FROM PHAEOPHYTES

Fucans from phaeophytes are a heterogeneous family of polysaccharides based on L-fucose ranging from fucose and sulphate-rich molecules to molecules less sulphated and richer in uronic residues. With alginates, these polysaccharides probably have specific functions in relation with the

| Order | Genus | Fucoidans | Ascophyllans | Sargassans |
|---|---|---|---|---|
| Estocarpales | *Ectocarpus* | | +(traces) | |
| | *Sorocarpus* | + | + | + |
| Chordariales | *Chorda* | ++ | | |
| | *Heterochordia* | ++ | | |
| | *Leathesia* | ++ | | |
| | *Nemacystus* | ++ | | |
| | *Spaerotrichia* | ++ | | |
| | *Tinocladia* | ++ | | |
| Desmarestiales | *Desmarestia* | + | + | ++ |
| Dictyosiphonales | *Asperococcus* | | | + |
| Scytosiphonales | *Colpomenia* | | | ++ |
| | *Scystosiphon* | + | + | + |
| Sphacelariales | *Stypocaulon* | | | |
| Dictyotales | *Dictyopteris* | + | ++ | |
| | *Dictyota* | | ++ | + |
| | *Padina* | + | ++ | + |
| | *Taonia* | | | + |
| Laminariales | *Alaria* (thalles) | | + | + |
| | *Alaria* (sporophylles) | | | ++ |
| | *Ecklonia* | + | | |
| | *Eisenia* | ++ | + | |
| | *Kjellmaniella* | ++ | ++ | |
| | *Laminaria* | ++ | | |
| | *Lessonia* | ++ | + | |
| | *Macrocystis* | ++ | | |
| | *Nereocystis* | ++ | | |
| | *Undaria (thalles)* | | + | + |
| | *Undaria* (sporophylles) | | | ++ |
| Fucales | *Ascophyllum* | + | ++ | + |
| | *Bifurcaria* | + | + | |
| | *Fucus* | + | + | |
| | *Halidrys* | + | | |
| | *Himanthalia* | + | + | |
| | *Hizikia* | | ++ | |
| | *Pelvetia* | ++ | + | |
| | *Sargassum* | | + | ++ |
| | *Turbinaria* | + | ++ | |

Table I : Diversity and relative abundance of the sulphated fucans from phaeophytes intertidal and marine environment [62,63]. Fucans are composed of three major species fucoidans, ascophyllans and glycuronofucoglycans also named sargassans. Briefly, they are extracted from the matrix phase of the cell walls of brown algae by organic solvents or water followed with aqueous treatments in presence of salts, acid, alkali or chelators [62-64]. All fucan species probably coexist in the thallus of the algae. The exact composition varies with algal species [62] (table I). Moreover for a given species the overall proportions of the constituents depend upon the tissue and numerous factors such as age, habitat or season [2,62] and these variations exert an effect upon the properties and qualities of the various extracts.

Fucoidans are described as primarily composed of α-1->2 and 1->3-4-sulphuryl-L-fucose with branching or sulphate at position 3 [65,66] (fig 2A). Pure sulphated fucoidans have been isolated [67]. Structurally, they are considered as highly charged and branched random coil [13,63,68]. The most recent structure of fucoidan from *Fucus vesiculosus* was proposed by Patankar *et al.* in 1993 [66]. In practice, fucoidans always contain small proportions of D-xylose, D-galactose, D-mannose and even L-arabinose as well as uronic acids. The average molar ratio of sulphate groups to total sugars (including uronic acid) are 3:2 suggesting that three moles of sulphate may be attached to two fucose residues in the polysaccharides [63,66,69]. However, it cannot be excluded that galactose or uronic acid residues might be sulphated. Most of the sulphate groups are estimated on the C-4 position of fucose [70].

Ascophyllans (fig. 2B) are considered as interfibrillar matrix polysaccharides which could play a role in the structure of the cell wall in the adult thallus by crosslinking alginates and cellulose microfibrils of the skeleton, while fucoidans are components of the intercellular matrix [62]. They are xylofucoglycuronans with large proportions of 4-sulphated L-fucose (α-1->2 linkages), D-xylose and uronic residues. Their structural hypothesis consists of a highly heteropolymeric structure with no long sequence of the same monosaccharide, xylose and glucuronic acids would then appear as 1->4 linked or terminal units [70,71].

Sargassans (fig 2C) are found in *Sargassum* [71] but also in the mature sporophylls of *Undaria* [72]. Their structure remains controversial, however many authors agree on linear chains of 1->4-linked D-galactose or glucose branched at C6 with L-fucosyl-3-sulphate or, occasionally, an uronic acid, probably D-glucuronic acid [63,73]. In general, galactose is found in the terminal position or 13 linked with other hexoses and, hence, does not form real homogalactan sequences [62].

The table II presents the average composition of the fucans which have been extracted from brown algae. Discrepancies revealed in the compositions might result from differences in the protocols used for the preparation and the fractionation of fucans. In addition to sugars and sulphate, each species can also contain a slight but variable amount of

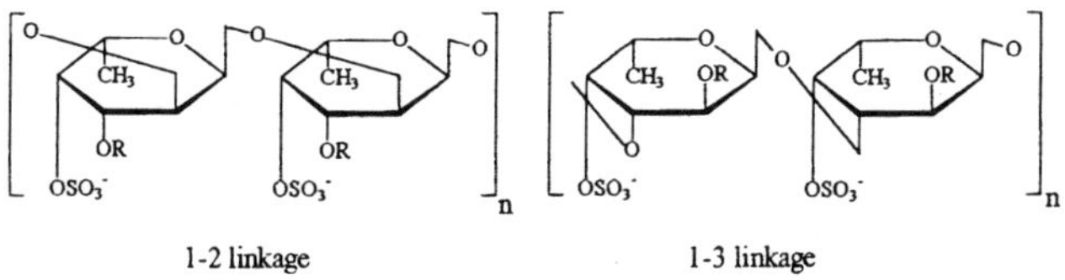


**(R=branching or sulphate)**

**A**

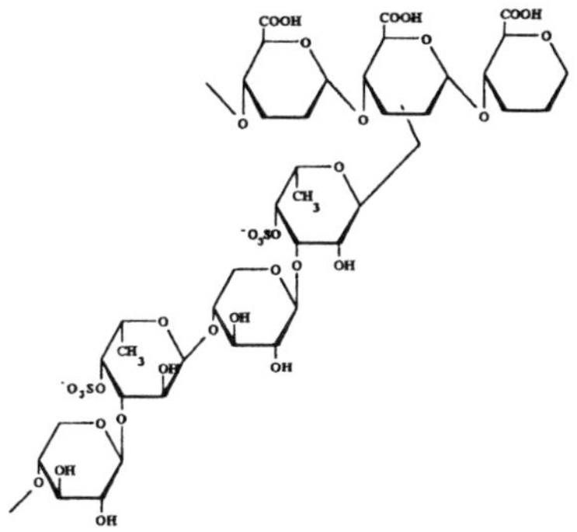

**B**

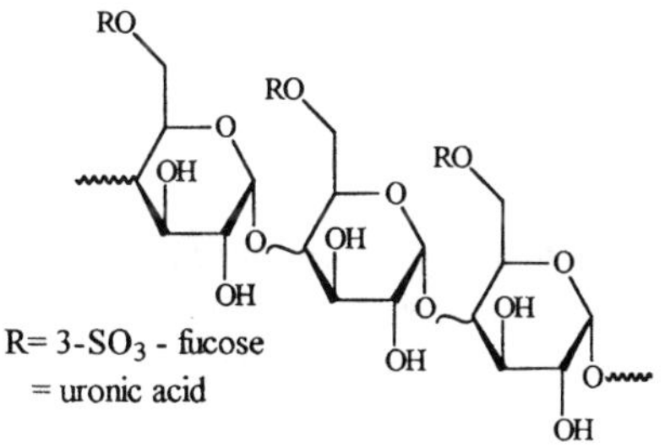


**C**

**Figure 2** : Schematic structure of fucaoidan (A), ascophyllan (B) and sargassan (C), from Kloareg and Quatrano [62].

nitrogen which probably reflects the presence of remaining protein fragments although Nishino *et al.* detected small amounts of glucosamine in commercial extracts from *Fucus vesiculosus* [74]. Nevertheless, in most cases what is described as one of the three species looks like a mixture. For example, if fucoidan is by far the most sulphated component of fucans, it is not the pure sulphated poly-L-fucose that is often quoted in the literature.

| | **L-fucose (g/100g)** | **Uronic acid (g/100g)** | **Other sugars (g/100g)** | **Sulfate groups (g/100g)** | **Proteins (g/100g)** |
|---|---|---|---|---|---|
| Fucoidans | **50-90** | **<8** | **5-40** | **35-45** | **<4** |
| Ascophyllans | **~25** | **~25** | **~25** | **~13** | **~12** |
| Sargassans | **25-45** | **~12** | **~36** | **15-21** | **4** |

Table II: Average chemical composition of the different fucan species (from Boisson-Vidal *et al.* [2])( From Kloareg and Quatrano [62] with permission of authors)

## 4. ANTICOAGULANT AND ANTITHROMBOTIC ACTIVITIES OF FUCANS FROM BROWN SEAWEEDS

In 1957, Springer *et al.* reported that fucoidans extracted from the brown marine alga *Fucus vesiculosus* possess anticoagulant activity *in vitro* and *in vivo*, which was neutralized by protamine sulphate showing the contribution of sulphate groups [15]. Abdel-Fattah *et al.* isolated a sulphated heteropolysaccharide from *Sargassum linifolum* (sargassan) composed of fucose, galactose, mannose, xylose, glucuronic acid, and a protein component, and demonstrated that the heteropolysaccharide possess higher anticoagulant activity than heparin [71]. Similar sulphated heteropolysaccharides having anticoagulant activity were also isolated from *Dyctyota dichotoma* [75], *Padina tetrastromatica* [76], *Padina pavonia* [77], *Eisenia bicyclis* (sulphated fucan) [78], *Undaria pinnatifida* (sulphated fucogalactan) [68], *Eklonia kurome* (sulphated galactofucan) [79], *Hizikia fusiforme* (sulphated galactofucan) [80], *Laminaria japonica* [92], *Ascophyllum nodosum* [82-84], *Pelvetia canaliculata* [85,86] and *Fucus vesiculosus* [15,67].

Fucoidans share certain properties with heparin and dermatan sulphate. However, the anticoagulant activities are not completely related to a heparin-like mechanism [68,71,78]. Fucoidans catalyze thrombin inhibition by AT and HCII but, unlike heparin, without strongly interacting with thrombin via AT [85,87]. In addition they were shown to interact directly with thrombin and with fibrinogen *in vitro* [67,79,82,88]. A high molecular weight commercially available fucoidan extracted from *Fucus vesiculosus* was described to exhibit anti-Factor Xa activity at very high concentration [87]. At the same time, Nishino *et al.* described fucoidan fractions from *Eklonia kurome* which catalyse the inhibition of thrombin mainly via HCII as does dermatan sulphate [89]. Soeda *et al.* have demonstrated that the commercial fucoidan and its derivatives (oversulphated and desulphated), unlike heparin, stimulated Tissue Plasminogen Activator, catalyzed plasminogen activation and prevented the formation of fibrin polymer by protecting plasmin activity from alpha-2-plasmin inhibitor. These fractions also decreased the rate of fibrin polymer formation, preventing clot formation by interfacing with fibrinogen. These activities increased with the degree of sulphation [90] and decreased with the molecular weight [91]. Moreover Durig *et al.* described a relatively low molecular weight fucoidan (50,000 g/mol), also prepared from commercial fucoidan, that combines potent anticoagulant activity and fibrinolytic properties, but with only minor platelet activating effects [92].

Up to 1994, no relationship between the structures of fucans and their biological activities have been given because of the lack of available informations on their detailed chemical. In 1994, B. Mulloy *et al.*[93] have shown that the anticoagulant activities of chemically modified fucans extracted from sea urchin was related to the presence of sulphated fucose. These fucans are composed of tetrasaccharide repeating units in which the 4 residues are 1,3 linked α-L fucopyranosyl units with different patterns of sulphation at positions 2 and 4. They present more homogeneous structures comparing to algal fucans. The same authors showed that the specific spatial array of the sulphated fucose branches was also essential for the anticoagulant property of a fucosylated chondroitin sulphate, a glycosaminoglycan extracted from the sea cucumber body wall [94]. They made the hypothesis that these groups may constituted the structural requirement for the binding of the polysaccharide to HCII and AT. In the case of a dermatan-sulphate like glycosaminoglycan extracted from ascidian, the sulphate groups at the 2-position of L-iduronic acid and 4-position of the N acetyl β-D galactosamine residues play also a key role in the recognition of the polysaccharide by HCII [95].

Clearly, like the activities described above, the anticoagulant activity of fucans depends upon the structure of the starting material, i.e. the structural heterogeneity, the sulphate pattern, and the molecular mass dispersion [63,67,79,88,96,97]. In our laboratory, we have also observed that the anticoagulant activity of fucans is very dependent on their chemical composition and molecular weight [2]. To assess relationships between fucan structure and activity, the anticoagulant properties of various fractions extracted from different algae species were studied [82,85]. Cell walls were isolated from the brown algae *Pelvetia canaliculata* (*Pc*), *Fucus vesiculosus* (*Fv*), *Sargassum muticum* (*Sm*), *Ascophyllum nodosum* (*An*), *Undaria pinatifida* (*Up*) *and Laminaria digitata* (*Ld*). Extracts from *Pc*, *Fv*, *Sm* and *Ld* were submitted to sequential chemical extraction. For each species four fractions were analyzed [2,98] and table III summerizes their anticoagulant activity. (All the abreviations that follow are defined in the legend of the table III.) The anticoagulant activity of the fucans was correlated to the chemical composition of the different fractions [82]. The TF, FF and HF fractions of *Pc* and *Fv* exhibit higher anticoagulant activity than those extracted from *Ld* and *Sm*. The former fractions were shown to have higher contents of L-fucose units and sulphate, and lower contents of uronic acid

| Algae fractions | FF | TF | HF | OHF | HF-CPC |
|---|---|---|---|---|---|
| *Pelvetia canaliculata* | 17 | 10 | 17 | 5 | 44 |
| *Fucus vesiculosus* | 19 | 5 | 19 | 12 | - |
| *Laminaria digitata* | 9 | - | 1 | 1 | - |
| *Sargassum muticum* | 10 | 0.5 | 5 | 2 | - |
| *Ascophyllum nodosum* | - | - | - | | 53 |

Table III: Specific anticoagulant activity of fucans expressed in equivalent International Units of heparin per mg of dry fucan (IU/mg) [2,82].
Standard heparin H108 activity a = 173 IU/mg. The chemical composition of the fucan fractions is described in detail in references [82,98]. FF, free fucans extracted with cold water; TF, fucans solubilized by Triton ; HF, fucans extracted with acidic water ; OHF, fucans extracted with alkaline water ; HF-CPC, fucans extracted with acidic water and purified with N-cetylpyridinium chloride.

and galactose, compared with those extracted from *Ld* and *Sm* [82,98]. Furthermore, within the same species, the overall chemical composition of fucans was related to the extraction conditions [98]. The general trend for the anticoagulant capacity among these fractions was HF-CPC>FF>HF>OHF> TF. The highest anticoagulant activity was found with the most sulphated fractions, HF-CPC, which exhibited an activity equivalent to 25 to 33% of that of heparin [2,82]. The anticoagulant activity of two crude fractions (HF-CPC from *An* and TF from *Pc*) was related to an heparin-like effect on AT and to a direct interaction with thrombin [2,82]. The effect on HCII was not tested. Each fraction was then sub-fractionated by low pressure steric exclusion chromatography. Studies on these sub-fractions ranging from 5 to 50 kD showed that the anticoagulant activity in Activated Partial

Thromboplastin Time (APTT) clotting assays, decreased with decreasing molecular weight [82]. With this study, Fucans from *An* were revealed as the most active.

In the same time, an anticoagulant LMW fucoidan has been isolated from *Pc*. The fraction had a molecular weight similar to that of standard heparin (~20,000g/mol) but was found twenty times less active than heparin [85]. However the fraction was also demonstrated to be as efficient as heparin and dermatan sulphate on HCII potentiation on a weight basis but 30 times less potent than heparin for action mediated by AT. Unlike heparin and commercial fucoidan studied by Church *et al.*, [87] no anti-Factor Xa activity was detected in the presence of fucoidan from *Pc* under the same experimental conditions. This can be partly explained by the low molecular weight of the fraction studied as compared to that of commercial fucoidan.

## 5. STRUCTURE – ANTICOAGULANT ACTIVITY RELATIONSHIPS OF FUCANS FROM *ASCOPHYLLUM NODOSUM*

We have prepared LMW fucans from *Ascophyllum nodosum* using acidic degradation, size exclusion chromatography, radical process depolymerization and ion exchange chromatography [86,99]. Then affinity chromatographies either on AT or on HCII of the most active fraction and structural studies have been performed.

**Acidic degradation**

The figure 3 summarizes the different conditions of preparation. A crude fucan extract (A) was obtained from *An* in a 1% (w/w) overall yield after treatments of the seaweed with organic solvents followed by acidic degradation in presence of calcium chloride to remove contaminating alginates. The extract was hydrolysed by dilute sulfuric acid to give the degraded acidic extract (HA). Then, HA was fractionated by low pressure SEC into three fractions (F1-F3) [86] or five fractions (G1-G5, H1-H5). A chromatogram of HA is given in figure 4. The distribution of molecular weights is very large and irregular, revealing heterogeneities in the size of the macromolecules. Fractionations were performed by following, more or less, the elution profile. Chemical characterizations and anticoagulant activities are summarized in table IV. The overall yields from the dried alga are much less than 1% (w:w) and descrepancies from one experiment to another lead to large variations in the standard deviations. Compositions and molecular weight both vary and it is difficult to make simple correlations with the anticoagulant activity, which remains low compared with heparin. However we can say that this activity varies with the sulfate to fucose ratio. Increasing in the temperature of the hydrolysis leads to the H fractions and

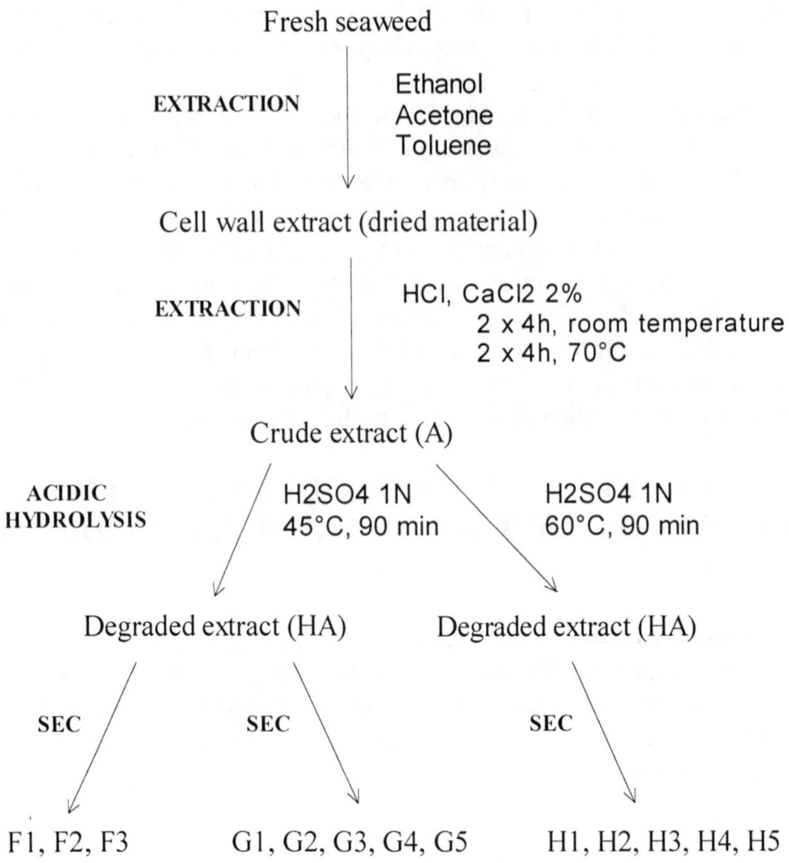


Fig. 3: Preparation schemes of fucan fractions by using acidic degradation followed with low pressure size exclusion chromatography (SEC).

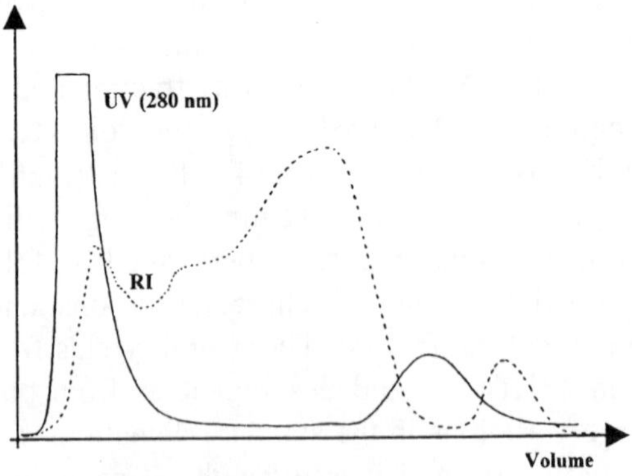


Fig. 4. Low pressure size exclusion chromatogram of HA extracted from *Ascophyllum nodosum* ($H_2SO_4$ 1N, 45°C, 90 min). Detections were performed by using refractive index (RI) and UV detectors (UV) [86].

| | Yields (g/100g) | | L-fucose | | Sulphate | | Uronic acid | | Mp (±3-7000) (g/mol) | N (g/100g) | APTT* (IU/mg) |
|---|---|---|---|---|---|---|---|---|---|---|---|
| | From dried alga | From A | g/100g | molar ratio per fucose | g/100g | Molar Ratio per fucose | g/100g | molar ratio per fucose | | | |
| A | 0.9 – 4.1 | - | 40 | 1.0 | 25 | 1.0 | 8 | 0.2 | 100,000 – 500,000 | 0.2 | 7 |
| HA | 0.8 – 4.1 | 80 – 100 | 65 | 1.0 | 25 | 0.6 | 9 | 0.1 | 4,000 – 500,000 | 0.3 | 7 |
| F1 | 0.60 | 27 | 33 | 1.0 | 21 | 1.0 | 16 | 0.4 | 70,000 – 500,000 | 0.2 | 10 |
| F2 | 0.80 | 35 | 44 | 1.0 | 28 | 1.0 | 7 | 0.1 | 20,000 | <0.1 | 8 |
| F3 | 0.11 | 6 | 36 | 1.0 | 27 | 1.2 | 5 | 0.1 | 10,000 | <0.1 | 8 |
| G1 | 0.15 | 0.13 | 39 | 1.0 | 22 | 0.9 | 8 | 0.2 | 81,000 | 0.1 | 9 |
| G2 | 0.11 | 0.10 | 42 | 1.0 | 21 | 0.8 | 6 | 0.1 | 47,000 | 0.1 | 8 |
| G3 | 0.13 | 0.11 | 41 | 1.0 | 30 | 1.2 | 4 | 0.1 | 27,000 | 0.2 | 8 |
| G4 | 0.15 | 0.13 | 38 | 1.0 | 27 | 1.1 | 4 | 0.1 | 13,000 | 0.1 | 5 |
| G5 | 0.08 | 0.07 | 39 | 1.0 | 28 | 1.2 | 1 | <0.1 | 13,000 | 0.1 | 6 |
| H1 | 0.20 | 28 | 13 | 1.0 | 20 | 2.4 | 19 | 1.2 | 94,000 | 0.3 | 4 |
| H2 | 0.08 | 12 | 23 | 1.0 | 29 | 2.0 | 9 | 0.3 | 24,000 | 0.2 | 6 |
| H3 | 0.14 | 19 | 28 | 1.0 | 38 | 2.2 | 6 | 0.2 | 14,000 | 0.2 | 4 |
| H4 | 0.24 | 30 | 30 | 1.0 | 47 | 2.5 | 3 | 0.1 | 8,000 | 0.2 | 2 |
| H5 | 0.11 | 12 | 24 | 1.0 | 30 | 2.0 | 3 | 0.1 | 6,000 | 0.2 | 1 |
| SD (%) | 20-80 | 20-80 | 10-40 | - | 10-30 | - | 10-30 | - | - | 10-30 | 10-40 |

Activated Partial Thromboplastin Time : standard international anticoagulant assay expressed in International Units per mg as compared to standard heparin : 173 IU/mg.

**Table IV** : Characterizations and anticoagulant activity of fucan fractions prepared by acidic degradation and size exclusion chromatography. Mp : chromatographic molecular weight. L-fucose content was obtained by using cystein-sulfuric colorimetric assay and from gas chromatography. Uronic acid content was obtained by using both metahydroxydiphenyl and carbazole colorimetric assays. The amounts of sulphate groups and protein contents were determined by using elemental analysis of respectively S and N. Molecular weights of fractions have been established by HPSEC using connected Lichrospher Si300 Diol (Merck) and hemaSecBio40 (Alltech) analytical columns at a flow rate of 0.5 ml/min with refractive index detection. Calibration was performed with standard pullulans and oligosaccharides (1.2-850 kD). Data were processed using Chromstar software (Merck-Bruker).

induces a strong decrease in the fucose content without modifications of the overall sulphate content. Moreover, the high sulphate to fucose molar ratio, often higher than 2, implies that H fractions are mixtures of different species. In all cases the remaining components were identified as neutral sugars, mainly xylose and, to lesser extent, galactose and mannose (data not shown). The chemical composition was in accordance with previous studies [70,100]. Electrophoresis performed on cellulose acetate (figure 5) have shown that all fractions were more or less mixtures of at least two species, an ascophyllan-like species (X) and a fucoidan-like species (Y). The decrease of the latter with the molecular weight can partly explained the variations in the uronic acid contents. Acidic hydrolyses of extracts from *An* followed by size exclusion chromatography did not then allow us to obtain a reproducible well defined anticoagulant fucan fraction.

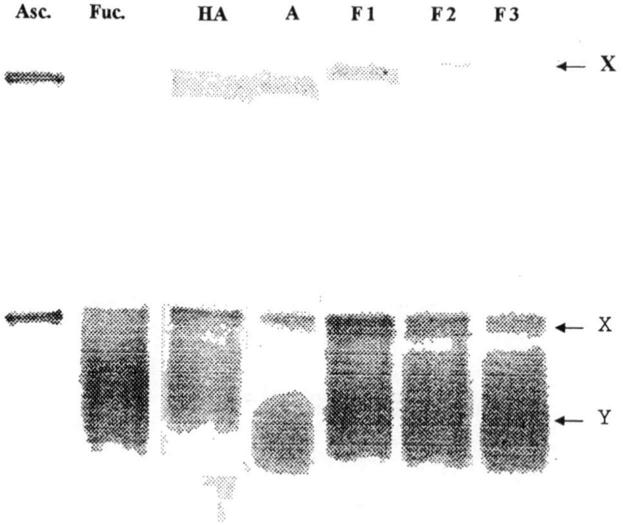


Fig. 5: Electrophoreses on cellulose acetate of a rude fucan extract (A), a degraded fucan (HA), F1, F2 and F3 as compared with commercial ascophyllan (X) and fucoidan (Y).

**Radical process degradation**

Studies performed on heparin and dermatan sulphate have shown the possibility to obtain in high yields low molecular weight fragments by using a radical process degradation system [101,102]. The reaction, named oxydative-reductive depolymerization, occurs through the formation of free hydroxyl radicals produced by the action of a metallic ion ($Cu^+$, $Fe^{2+}$) with the hydrogen peroxide ($H_2O_2$) or a peracid, at well defined temperature and pH [103]. At neutral pH and 60°C, the reactive species degrade

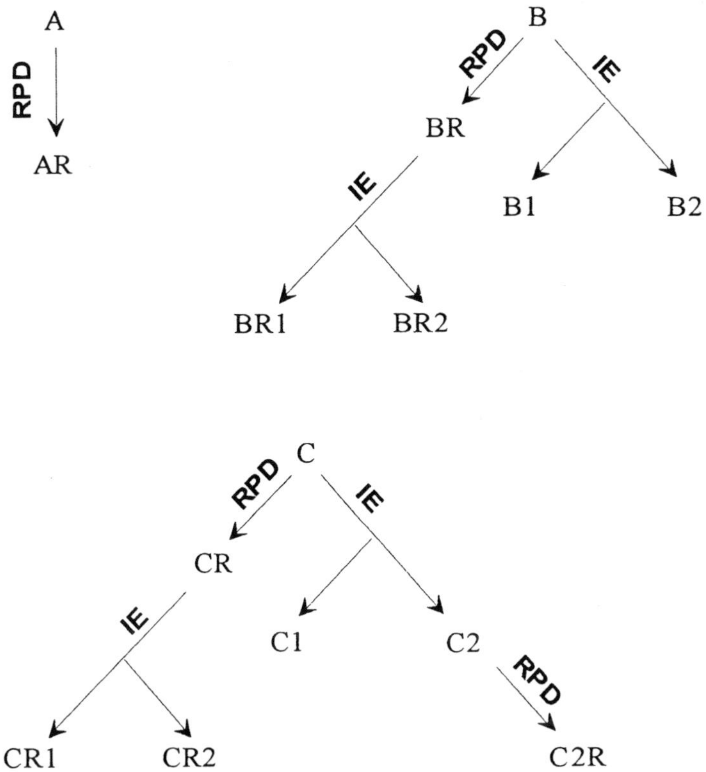


Fig. 6. Preparation schemes of fucan fractions from crude extracts A, B and C by using radical process degradation (RPD) and ion exchange chromatography (IE)

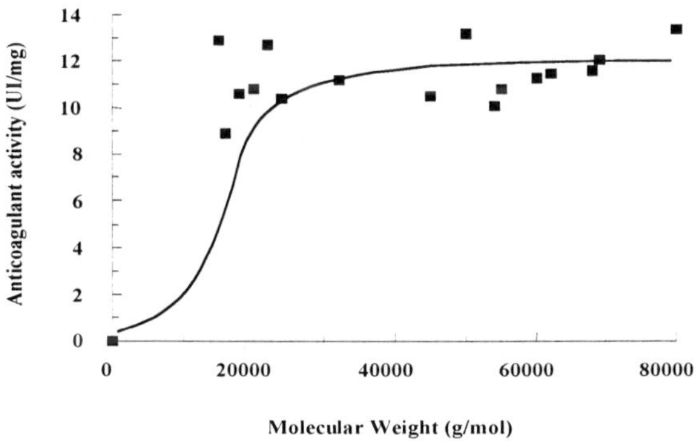


Fig. 7. Anticoagulant activity – Molecular weight relationship of fucan fractions from *Ascophyllum nodosum* prepared by radical process degradation () and acidic degradation () and presenting the same average chemical composition.

polysaccharides more efficiently than the classical acidic hydrolysis without epimerizations or openings of glycosidic cycles [101,104].

LMW fucans were prepared by radical process degradation from an acidic extract from *An* [97,99]. Two protocols have been used (protocol 1 and 2), the protocol 2 being more drastic in terms of temperature of reaction and concentrations of chemicals. In this case, the radical process degradation proceeds through the formation of free radicals from the hydrogen peroxide-cupric redox system [105]. From a HMW fucan of 80,000 g/mol, the protocol 1 allowed us to obtain molecular weights around 20,000 g/mol in about 3 h in a 50% yield without modification of the average chemical composition. Longer times of reaction only led to a decrease in the yield. Several fractions with molecular weight ranging from 20,000 g/mol to 80,000 g/mol have been prepared. We have plotted anticoagulant activity - molecular weight relationship, by taking into account three fractions with molecular weight lower than 20,000 g/mol obtained from the previous acidic degradation - fractionation protocol and presenting the same average chemical composition as the fractions obtained by radical process degradation. The anticoagulant activity increased up to 20,000 g/mol and remained constant for higher molecular weights (fig 7).

In order to obtain fractions with molecular weights lower than 20,000 g/mol we performed degradation under more drastic conditions (protocol 2). Three series of fractions were prepared from three HMW fucans: A (Mp = 556,000 g/mol), B (Mp = 100,000 g/mol and 516,000 g/mol) and C (320,000 g/mol). The figure 6 displays the scheme of all experiments that have been performed. Compositions, molecular weight and APTT activities of all fractions are presented in Table V. The degradation of A was monitored during the degradation process and analyzed with HPSEC. After 30 min, a 66,000 g/mol fucan fraction peak with a 19,000 g/mol shoulder was obtained. Between 1 h and 5 h, the molecular weight decreased to give AR in a yield of 47% and presenting a single peak at 8,300 g/mol (AR). The degradation process was similar for B and C (data not shown), leading respectively to BR and CR in similar yields. The HPSEC chromatograms show that degraded fractions present a more narrow molecular weight distribution compared with the parent compounds [97]. However, the average chemical compositions of the LMW fractions were different from that of the starting fucan. Moreover, the charged groups compositions of the former were different from that of the latter as reflected by electrophoreses shown in the figure 8. We have observed that the starting fucans were mixtures of at least a fucoidan-like and an ascophyllan-like species, the radical process degrading mainly the latter. Low pressure ion-exchange chromatography experiments allowed us to separate both species [97]. For example, fractions B1 and BR1, collected with a gradient of 0 to 0.75 M sodium chloride, showed a high amount of D-glucuronic acid as compared to B and a decrease in the L-fucose and sulphate contents. Their APTT

| Fraction | Yield (%) | L-fucose (g/100g) | $SO_3Na$ (g/100g) | D-glucuronic Acid (g/100g) | Mp ( ±3 – 7,000) (g/100g) | APTT (IU/mg) |
|---|---|---|---|---|---|---|
| A | - | 31 | 26 | 6 | 556,000 | 9 |
| B | - | 36 | 18 | 12 | 516,000[a] | 12 |
| C | - | 38 | 30 | 6 | 320,000 | 14 |
| AR | 47 | 36 | 30 | 3 | 8,300 | 8 |
| BR | 50 | 32 | 30 | 7 | 7,800 | 7 |
| CR | 50 | 31 | 24 | 1 | 7,000 | 5 |
| B1 | 30 | 23 | 14 | 24 | 156,000 | 2 |
| B2 | 40 | 43 | 35 | 2 | 600,000[a] | 25 |
| BR1 | 13 | 23 | 20 | 18 | 6,000 | <1 |
| BR2 | 50 | 43 | 35 | 1 | 13,000 | 8 |
| C1 | 22 | 34 | 20 | 17 | 450,000 | 5 |
| C2 | 36 | 59 | 45 | 2 | 87,000 | 16 |
| C2R | 11 | 63 | 45 | 1 | 15,000 | 10 |
| CR1 | 10 | 6[b] | 13 | 2 | <3,000 | 1 |
| CR2 | 40 | 57 | 36 | 1 | 7,000 | 11 |
| SD | 5-10% | 5-10% | 5-10% | 5-10% | - | 10-20% |

[a] shoulder at 100,000 g/mol
[b] xylose, galactose and mannose were also detected in high amounts by GC (data not shown).

Table V. Characterization of fucan fractions. (see Table IV for experimental conditions).

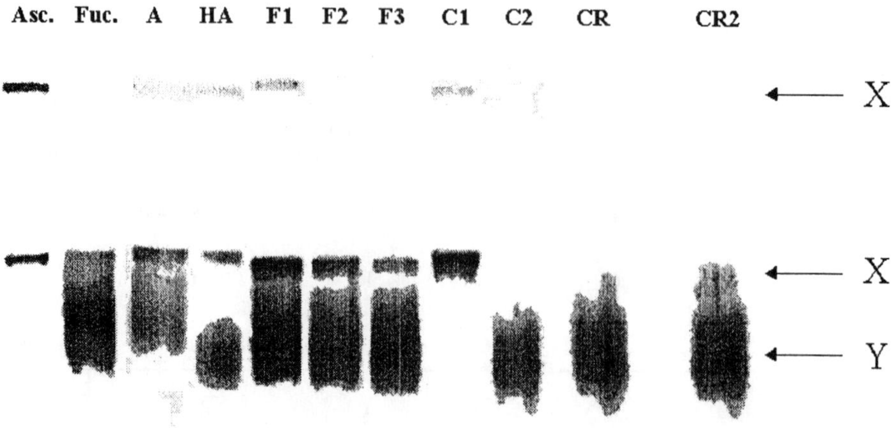


Fig. 8. Electrophoreses of fucan fractions obtained by radical process degradation and ion exchange chromatography as compared to commercial ascophyllan (X) and fucoidan (Y) species.

activities were much lower as compared to B. In contrast, B2 and BR2, eluted above 0.75 M sodium chloride were enriched with sulphate and L-fucose while their D-glucuronic acid content was five fold lower than that of the parent compounds. Similar results have been obtained with A and C (data not shown). In all cases, the molecular weight of the ascophyllan-like fraction (1) was found much higher than that of the fucoidan-like one (2) and the radical process degradation seemed to degrade more efficiently the former. Nevertheless, only fucoidan-like species were responsible for the anticoagulant activity of the parent compound.

Finally, it was more efficient to prepare anticoagulant LMW fucan by performing radical process degradation followed by ion-exchange chromatography. The low standard deviations implied a quite good reproducibility within each series of experiments, compared with the acidic degradation-fractionation process.

**Affinity chromatography of CR2 with HCII and AT**

The anticoagulant activity of fucan is attributed mainly to its capacity as catalyst of the thrombin-HCII complex formation and, to a lesser extent, of the thrombin-AT complex formation. The affinity chromatography of fucan fractions on HCII and AT may allow, as a final step, to isolate the anticoagulant macromolecular species.

The CR2 fraction was mixed with AT or HCII and the obtained complex was immobilized on a Sepharose-concanavalin A gel following the protocol used by Lam *et al.* with heparin [106] and Sinninger *et al.* with dermatan sulphate [107]. The elution profiles allow to separate in each case two fractions (fig 9). The high affinity fraction on AT is eluted with an ionic strenght of about 0.1 M NaCl i.e. five times lower that the one used for the high affinity fraction on HCII. This confirms the selectivity of fucoidan for HCII. The chemical characterizations and the anticoagulant activity of the fractions are gathered in table VI. With the highest anticoagulant activities, the most retained fractions (H-AT and H-HCII) also exihibit the highest sulphate to fucose ratios. However, we cannot assign the affinity of the CR2 fraction only to ionic interactions. Indeed, this fraction is the most retained in ion exchange chromatography and we make the hypothesis that some structural features exist in the fucan fractions interacting specifically with the inhibitors.

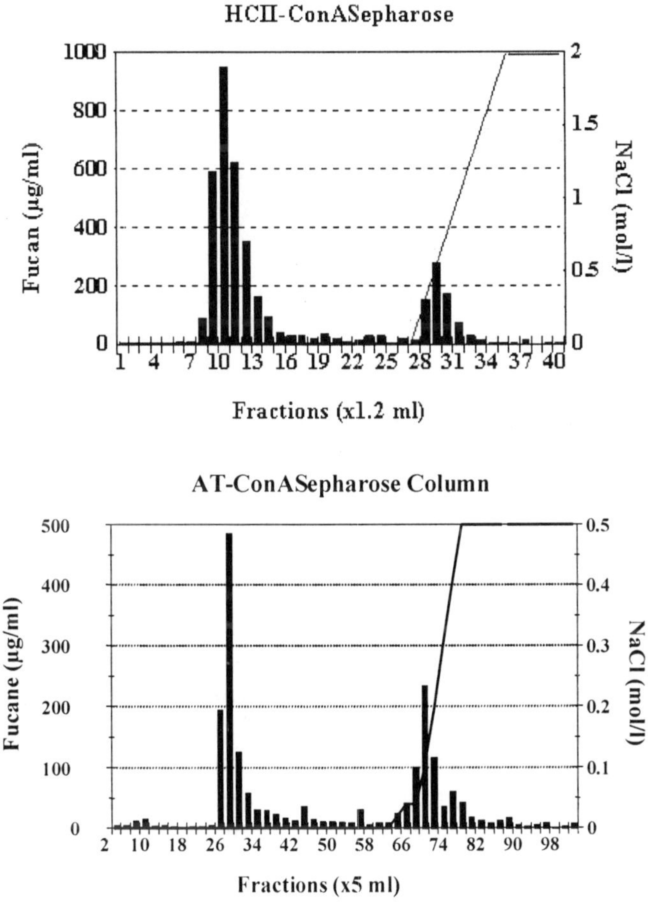


Fig. 9. Affinity chromatograms of the CR2 fucan fraction on HCII

| | L-fucose | | Sulphate | | Uronic acid (g/100g) | Specific Anticoagulant activity* (µg/ml) |
|---|---|---|---|---|---|---|
| | g/100g | Molar ratio per fucose | g/100g | Molar ratio per fucose | | |
| CR2 | 57 | 1.0 | 36 | 1.0 | 1 | 25 |
| L-AT | ND | - | 28 | - | ND | ND |
| H-AT | 44 | 1.0 | 33 | 1.2 | <1 | 15 |
| L-HCII | 54 | 1.0 | 33 | 1.0 | 2 | ND |
| H-HCII | 42 | 1 | 35 | 1.3 | 2 | 10 |

* expressed as the concentration of tested compound which doubles a standard coagulation time of 40s. ND: not determined.

Table VI: Characterizations and anticoagulant activity of fucan fractions from CR2 with low and high affinity with AT and HCII. (See table IV for experimental conditions).

**Spectroscopic studies by FTIR and high field n.m.r**

IR spectra of the fucan fractions C, CR, CR1 and CR2 were presenting the same pattern correponding to fucose rich and sulphate rich polysaccharides; mainly S=O stretching between 1230 and 1260 $cm^{-1}$, C-O-S bending at 820 $cm^{-1}$ (equatorial position) and at 845 $cm^{-1}$ (axial position) and C-C-H bending of methyl group in L-fucose at 1380 $cm^{-1}$ [108-110]. The main differences were found in the 1800-600 $cm^{-1}$ region. Several bands between 1410 and 1450 $cm^{-1}$ (C=O stretching carboxylate) strongly decreased from C to CR, disappeared in CR2 and increased in the ascophyllan like fraction CR1 in accordance with the variation in the uronic acid content [111]. By the way the broad absorption at 1630 $cm^{-1}$ (strongly retained water molecules) was also shifted to 1619 $cm^{-1}$ (overlap with C=O carboxylate stretching) in the CR1 fraction. Finally, the vibrations at 845 $cm^{-1}$ strongly increased from C to CR2 while the vibration at 820 $cm^{-1}$ remained unchanged. The same observations were done with the fractions B1 and B2 as compared to B. Some descrepancies in the intensities of the bands at 845 $cm^{-1}$ between B1 and CR1 led us to conclude that the radical process degradation can modify the proportion of the axial and equatorial conformers of sulphate groups on glycosidic rings. On another hand, equatorial sulphate are mainly found on the C2 and C3 positions while axial sulphate are on the C4 position. In the ascophyllan like species sulphated fucosyl residues should be linked in 12. On the contrary, linkages should be 13 in fucoidan like species [66].

| | Hydrogen | H1 | H2 | H3 | H4 | H5 | H6 |
|---|---|---|---|---|---|---|---|
| Mulloy *et al.* [93] Ribeiro *et al.* [112] | unsubstituted | 5,3-5,6 | 3,9-4,0 | 4,0-4,4 | 3,9-4,1 | 4,3-4,5 | 1,2-1,3 |
| Pavao *et al.* [113] | Substituted ($SO_3Na$) | - | 4,5-4,6 | - | 4,7-4,9 | - | - |
| Data from CR2 and affinity fractions | unsubstituted | 5,2-5,6 | 4,0-4,1 | 4,2-4,3 | 4,0 | 4,3-4,5 | 1,2-1,4 |
| | substituted ($SO_3Na$) | - | 4,6-4,7 | - | 4,8-4,9 | - | - |

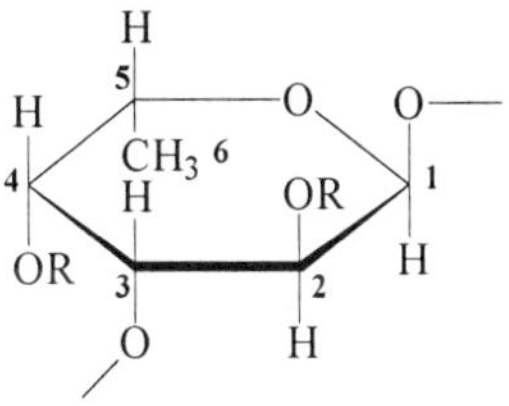


Fig. 10: n.m.r. $^1$H chemical shifts and the corresponding base fucosyl unit.

Due to the very low amount of the H-HCII fraction, only H-AT, CR2 and the low affinity fractions L-AT and L-HCII have been examined by $^1$H and $^{13}$C n.m.r. The 1D spectra of the fractions are similar although very complex. 2D COSY and HETEROCOSY experiments have been performed and the table on the figure 10 compares our data with that from linear fucoidan-like species extracted from echinoderms and from *Fucus vesiculosus* obtained by Mulloy *et al.*, Ribeiro *et al.* and Pavao *et al.* [93,112,113]. The spectrum is divided into three regions (Fig.10). Signals at 1.2 and 1.4 ppm corresponds to the methyl groups of fucose (H6). Anomeric H are found at low fields from 5.2 to 5.6 ppm (H1) with  configuration. Large overlapped signals between 3.8 and 5.0 ppm are attributed to H in the glycosidic cycles (H2-H5). Hydrogens born by substituted C2 and C4 are found at 4.65 and 4.80 ppm respectively, signals of unsubstituted cycles beeing observed below 4.5 ppm. Moreover, the difference of 0.7 ppm between H belonging to substituted rings (C2 and C4) and those from unsubstituted rings allows to define the substituants as sulphate groups. As a consequence, the glycosidic linkages between fucosyl residues are mainly 13.

Finally spectroscopic studies suggest that fractions with high affinity for AT and HCII are structures of about 15 to 20 sulphated  L-fucosyl units linked in 13 with mainly sulphate as branchings. Nevertheless further studies should have to clarify if these fractions are linear or branched structures.

## 6. CONCLUSION

For the last 50 years numerous seaweed extracts have been reported to exhibit some anticoagulant activity. In all cases sulphated polysaccharides were related to this activity. The purification of bioactive structure is clearly depending on the extraction process from the appropriate source material. We consider that there are as many methods for the preparation of extracts as there are extracts. However some general procedures stand out (for example the use of organic solvents and strong acids for extractions). In terms of structural characterization, discrepancies in the reports can be partly attributed to the wide variety of physico-chemical techniques used over the years. Moreover, as stated by Percival [114] in 1979: " the earlier studies on the structure of the polysaccharides from seaweeds were carried out before the emergence of modern and efficient physico-chemical techniques such as chromatography and NMR whose would have to allow more precise structural characterizations." In order to clarify the situation, it is of interest to prepare and to purify low molecular weight fucan fractions compatible with spectroscopic studies, in good yields and in a reproducible way. The use of a radical process degradation system seems to be the most convenient for this purpose with the ion exchange chromatography as an improvement. Moreover its efficiency and reliability provide room for further industrial developments.

In general anticoagulant fucan extracts from brown algae exhibit very complex chemical structures, with numerous different carbohydrates other than fucose and glucuronic acid, different levels of branchings and large molecular weight distribution. These extracts are often mixtures of the different fucan species and mainly fucoidan are found to be responsible for the anticoagulant activity. This fraction slows down the clotting of blood as heparin does, and we want to identify what features of its molecular structure cause this resemblance. In other words, do specific carbohydrate structures responsible for the anticoagulant activity exist in fucoidan ? Until now, only heparin has been found to present a unique carbohydrate structure endowing the polysaccharide with a strong anticoagulant capacity. On the other hand, the behaviour of fucan in contact with the blood coagulation seems to be closer to that of dermatan sulphate than to that of heparin. However the hypothesis of specific carbohydrate structures is required to explain the affinity of fucoidan for the blood coagulation inhibitor HCII, although weaker interactions with AT occur as well as the direct antithrombin effect. Thus we consider that the effect of fucan on blood coagulation is original. We suggest, as do McLellan and Jurd [3], that, in terms of biochemical mechanisms and as a general approach, fucan should be more realistically compared to dermatan sulphate although many studies still have to be performed on fucan interacting with blood coagulation factors. Finally, well characterized structures of anticoagulant fucan could serve as models for novel anticoagulant or antithrombotic drugs. Let us imagine fucan-like compounds with the advantages of heparin but without the drawbacks.

## ACKNOWLEDGMENTS

This work was supported by the Centre National de la Recherche Scientifique (CNRS) and by the Institut Français de Recherche pour l'Exploitation de la Mer (IFREMER). Some of the work on extracts from *An* was a part of the thesis of Alain Nardella. We specially aknowledge C. Sinquin for the preparation of the fucan fractions. The readership of Pr. Roger Marchand is also gratefully acknowledged.

## References

[1] Boisson-Vidal, C., Haroun, F., Ellouali, M., Blondin, C., Fischer, A-M., de Agostini, A. and Jozefonvicz, J. (1995), *Drugs of the Future*, J.R.Prous Ed., Prous Science, Barcelona, Philadelphia, 20:12. 1237-1249.

[2] Boisson-Vidal, C., Colliec-Jouault, S., Fischer, A-.MÊ., Tapon-Bretaudière, J., Sternberg, C., Durand, P. and Jozefonvicz, J. (1991), *Drugs of the Future*, J.R.Prous Ed., Prous Science, Barcelona, Philadelphia, 16:6. 539-545.

[3] McLellan, D.S. and Jurd, K.M. (1992), *Blood Coagulation and Fibrinolysis*, 3. 69-77.

[4] Casu, B., Oreste, P., Torri, G., Zoppetti, G., Choay, J., Lormeau, J.C., Petitou, M. and Sinay, P. (1981), *Biochem.J.*, 197. 599-609.

[5] Rosenberg, R.D. and Lam L.H (1979), *Proc.Natl.Acad.Sci.USA*, 76. 1218-1222.

[6] Lindahl, U., Backstrom, G., Thunberg, L. and Leder, I.G (1980), *Proc .Natl .Acad Sci.USA*, 77 6551-6555.

[7] Chargaff, E., Bancroft, F.W. and Stanley-Brown, M (1936),*J.Biol.Chem.*,115. 151-161.

[8] Elsner, H., Broser, W. and Burgel, E (1937), *Ztschr.Physiol.Chem.*, 246. 244-247.

[9] Deacon-Smith, R.A.,Lee-Potter, J.P.and Rogers, D.J(1985),*Botanica Mar.*, 28 333-338.

[10] Jurd, K.M., Rogers, D.J., Blunden, G. and McLellan, D.S(1995), *J.Appl.Phycol.*, 7. 339-345.

[11] Uehara, T., Takeshita, M. and Maeda, M(1992), *Carbohydr.Res.*, 235. 309-311.

[12] Kylin, H. (1913), *Hoppe-Seyler's Z.Physiol.Chem.*, 83. 171-197.

[13] Doner, L.W. and Whistler, R.L. . (1972), In "Industrial Gums" Ed. R.L.Whistler and J.N.BeMiller, Academic Press Inc 115-121.

[14] Percival, E.G.V. and Ross, A.G.: (1950) *J.Chem.Soc.*, 717-720.

[15] Springer, G.F., Wurzel, H.A., McNeal, G.M., Ansell, N.J. and Doughty, M.F. .(1957), *Proc.Soc.Exp.Biol.Med.*, 94. 404-409.

[16] McLean, J(1959), *Circulation*, 19 .75-78.

[17] Verstraete, M(1990), *Drugs*, 40. 498-530.

[18] Casu, B(1989), In "Heparin.Chemical and biological properties. Clinical applications. " Ed.D.A.Lane and U.Lindahl, Edward Arnold, London. 25-49.

[19] Bourin, M.C. and Lindahl, U(1993), *Biochem.J.*, 293. 313-330.

[20] Tollefsen, D.M., Majerus, D.W.and Blank, M.K. (1982) *J.Biol.Chem.*, 257.,2162-2169.

[21] Bjork, I. and Lindahl, U. (1982), *Mol.Cell.Biochem.*, 48. 161-182.

[22] Travis, J. and Salvesen, G(1983),.S*Annu.Rev.Biochem.*, 52. 655- 709.

[23] Tollefsen, D.M. and Blank M.K .(1981) *J.Clin.Invest.*, 68, 589- 596.
[24] Choay, J., Petitou, M., Lormeau, J.C., Sinay, P., Casu, B. and Gatti, G. . (1983), *Biochem.Biophys.Res.Commun.*, 116492-499.
[25] Rosenberg, R.D. and Damus, P.S. (1973) *J.Biol.Chem.*, 248, 6490-6505.
[26] Thunberg, L., Backstrom, G., and Lindahl, U(1982), *Carbohydr.Res.*, 100. 393-410.
[27] Rosenberg, R.D., Jordan, R.E., Favreau, L.V. and Lam, L.H (1979), *Biochem. Biophys Res.Commun.*, 86. 1319-1324.
[28] Hirsh, J., Ofosu, F.A. and Levine, M.N (1987), In "Thrombosis and haemostasis" Leuven University Press, Leuven 325-331.
[29] Holmer, H. (1989), In "Heparin.Chemical and biological properties.Clinical applications. " Ed.D.A.Lane and U.Lindahl, Edward Arnold, London. 575-596.
[30] Hirsh, J. and Levine, M.N. . (1992), *Blood*, 791-17.
[31] Godal, H.C., (1989), In "Heparin.Chemical and biological properties .Clinical applications. " Ed.D.A.Lane and U.Lindahl, Edward Arnold, London. 533-548.
[32] Levine, M.N., Hirsh, J. and Kelton, J.G(1989), In "Heparin.Chemical and biological properties.Clinical applications. " Ed.D.A.Lane and U.Lindahl, Edward Arnold, London. 517-532.
[33] Kindness, G., Williamson, F.B. and Long, W.F (1980), *Biochem.Soc.Trans.*, 8. 82-87.
[34] Andersson, L.O., Hoffman, J., Holmer, E., Larm, O., Larsson, K. and Söderström, G. (1982), "*Thromb.Res.*, 28. 741-747.
[35] Oshima, G., Nagai, T. and Nagasawa, K (1984), *Thromb.Res.*, 35. 601-611.
[36] Fischer, A-M., Mauzac, M., Tapon-Bretaudiere, J. and Jozefonvicz, J. (1987), *Biomaterials*, 6. 198-202.
[37] Chaubet, F., Champion, J., Maïga, O., Mauray, S. and Jozefonvicz, J (1995), *Carbohydr.Polym.*, 28. 145-152.
[38] Conrad, H.E. . (1989), *Ann.N.Y.Acad.Sci.*, 55618-28.
[39] Gallagher, J.T. and Lyon, M (1989), In "Heparin .Chemical and biological properties, Clinical applications. "Ed.D.A.Lane and U.Lindahl, Edward Arnold, London. 135-158.
[40] Teien, A.N., Abildgaard, U. and Hook, M. (1976) *Thromb.Res.*8. 859-867. "
[41] Ofosu, F.A., Modi, G.J., Blajchman, M.A., Buchanan, M.R. and Johnson, E.A. (1987), *Biochem.J.*, 248. 889-896.
[42] Hoppensteadt, D., Walenga, J.M. and Fareed, J. (1990), *Thromb.Res.*, 60. 191-200.
[43] Mascellani, G., Liverani, L., Bianchini, P., Parma, B., Torri, G., Bisio, A., Guerrini, M. and Casu, B. (1993), *Biochem.J.*, 296639-648.
[44] Agnelli, G., Cosmi, B., Di Filippo, P., Ranucci, V., Veschi, F., Longetti, M., Renga, C., Barzi, F., Gianese, F. and Lupatelli, L. (1992),. *Thromb.Haemost.*, 67203-208.
[45] Scully, M.F., Ellis, V. and Kakkar, V.V. (1986), *Thromb.Res.*, 41. 489-499.
[46] Dunn, F., Soria, J., Soria, C., Thomaidis, A., Tobelem, G. and Caen, P (1983), *Thromb. Res.*, 29. 141-148.
[47] Blondin, C., Fischer, E., Boisson-Vidal, C., Kazatchkine, M. and Jozefonvicz, J. . (1994), *Mol.Immunol.*, 31245-253.
[48] Blondin, C., Chaubet, F., Nardella, A., Sinquin, C. and Jozefonvicz, J (1996), *Biomaterials*, 17. 597-603.
[49] Baba, M., Snoeck, R, Pauwels, R. and De Clercq, E (1988), *Antimicrob.Agents Chemother.*, 32. 1742-1745.
[50] Nakashima, H., Kido, Y., Kobayashi, N., Motoki, Y., Neushul, M. and Yamamoto, N. (1987), *Antimicrob.Agents Chemother.*, 31. 1524-1528.
[51] Hanenberger, R. and Jakobson, A.M . (1991), *Glycoconjugate J.*, 8350-353.
[52] Glabe, C.G., Yednock, T. and Rosen, S.D(1983), *J.Cell Sci.*, 61. 475-490.
[53] Soeda, S., Ishida, S., Honda, O., Shimeno, H. and Nagamatsu, A (1994), *Cancer Lett.*, 85. 133-138.

[54] Belford, D.A., Hendry, I.A. and Parish, C.R(1993), *J.Cell.Physiol.*, 157. 184-189.
[55] Logeart, D., Prigent-Richard, S., Jozefonvicz, J. and Letourneur, D (1997), *Eur .J.Cell. Biol.*, 74. 376-384.
[56] Coombe, D.R., Parish, C.R., Ramshaw, I.A. and Snowden, J.M (1987), *Int.J.Cancer*, 39. 82-90.
[57] Ellouali, M., Boisson-Vidal, C. and Jozefonvicz, J (1994), *Colloids Surf.B : Biointerfaces*, 2. 305-314.
[58] Ellouali, M., Boisson-Vidal, C., Durand, P.. and Jozefonvicz, J (1991), *Anticancer Res.*, 13. 2011-2020.
[59] Huang, T.T.F. and Yanagimachi, R(1984), *Exp.Cell Res.*, 153. 363-373.
[60] Mahony, M.C., Clark, G.F., Oehninger, S., Acosta, A.A. and Hodgen, G.D (1993), *Contraception*, 48. 277-289.
[61] Mahony, M.C., Clark, G.F., Oehninger, S., Acosta, A.A. and Hodgen, G.D. .(1991), *Contraception*, 44 657-665.
[62] Kloareg, B. and Quatrano, R.S. . (1988), *Oceanogr.Mar.Biol.Ann.Rev.*, 26259-315.
[63] Kloareg, B., Demarty, M. and Mabeau, S . (1986), *Int.J.Biol.Macromol.*, 8380-386.
[64] De Reviers, B., Mabeau, S.et Kloareg, B. (1983), *Cryptogamie Algologie*, IV:1-. IV, 55-62.
[65] Medcalf, D.G. (1978), In "ÊCarbohydrate sulphatesÊ" Schweiger, R.G. (Ed) Am.Chem.Soc., 77. 225-244.
[66] Patankar, M.S., Oehninger, S., Barnett, T., Williams, R.L. and Clark, G.F .(1993), *J.Biol.Chem.*, 268:29 21770-21776.
[67] Bernardi, G. and Springer, G.F. . (1962), *J.Biol.Chem.*, 237:175-80.
[68] Mori, H., Kamei, H., Nishide, E. and Nisizawa, K .(1982), In "ÊMarine Algae in Pharmaceutical ScienceÊ" Hoppe, H.A. and T.Levring (Eds).W.de Gruyter Berlin, N.Y., 2 109-121.
[69] Anno, K.,Terahata, H., Hayashi, and Seno, N(1966),*Agr.Biol.Chem.Tokyo*.,30.495-499.
[70] Larsen, B., Haug, A. and Painter, T.J (1966), *Acta Chem.Scand.*, 20. 219-230.
[71] Abdel-Fattah A., F., Hussein, M.M.D. and Salem, H.M (1974),*Carbohyd.Res.*, 33 9-17.
[72] Igarashi, O., Iwaki, E. and Fukuka, H(1971), *Agr.Biol.Chem.*, 35:12. 1836-1843.
[73] Medcalf, D.G., Schneider, T.L. and Barnett, R.W (1978), *Carbohydr.Res.*, 66. 167-171.
[74] Nishino,T, Nishioka, C., Ura, H.and Nagumo, T. (1994), *Carbohydr.Res.*,255.213-224.
[75] Abdel-Fattah A.F., Hussein, M.M.D. and Fouad,T (1978),*Phytochemistry*, 17. 741-743.
[76] Prasada Rao, N.V.S.A.V., Sastry, K.V. and Venkata Rao, E (1984) *Phytochemistry* 2531-2533.
[77] Hussein, M.M.–D., Abdel-Aziz, A and Salem, H.M (1980), *Phytochemistry*, 19. 2131-2132.
[78] Usui, T., Asari, K. and Mizuno, T. *(*1980), *Agric.Biol.Chem*1965-1966.
[79] Nishino, T., Yokoyama G., Dobashi, K., Fujihara, M. and Nagumo T (1989), *Carbohyd.Res.*, 186119-129.
[80] Dobashi, K., Nishino, T, Fujihara, M. and Nagumo, T (1989), *Carbohydr.Res.*, 194. 315-320.
[81] Rozkin, M.Y., Levina, M.N., Kameneva, N.S., Usov, A.I. and Yefimov, V.S. (1989), *Farmak.i Toksikol.*, 3. pp 48-51.
[82] Grauffel, V., Kloareg, B., Mabeau, S., Durand, P. and Jozefonvicz, J. (1989), *Biomaterials*, 10. 363-368.
[83] Mauray, S., Strenberg, C., Thévenieaux, J., Millet, J., Sinquin, C, Tapon-Bretaudière, J. and Fisher A-M (1995), *Thromb.Haemost.*, 74:5. 1280-1285.

[84] Mauray, S., de Raucourt, E., Chaubet, F., Maïga-Revel, O., Sternberg, C. and Fischer, A-M. .(1998), *J.Biomater.Sci.Polymer.Edn.*, 9:4 373-387.

[85] Colliec, S., Fischer, A-M., Tapon-Bretaudière, J., Boisson, C., Durand, P. and Jozefonvicz, J. (1991), *Thromb. Res.*, 143-154.

[86] Colliec, S., Boisson-Vidal, C. and Jozefonvicz, J (1994), *Phytochemistry*, 35:3. 697-700.

[87] Church, F.C., Meade, J.B., Treanor, R.E. and Whinna, H.C . (1989), *J.Biol.Chem.*, 264. 3618-3623.

[88] Nishino, T. and Nagumo, T. . (1991), *Carbohydr.Res.*, 214193-197.

[89] Nishino, T., Aizu, Y. and Nagumo, T(1991), *Thromb.Res.*, 62. 765-773.

[90] Soeda, S., Sakaguchi, S., Shimeno, H. and Nagamatsu, A (1992), *Biochem. Pharmacol.*, 43:8. 1853-1858.

[91] Soeda, S., Ohmagari, Y., Shimeno, H. and Nagamatsu, A (1993), *Thromb.Haemost.*, 72. 247-256.

[92] Dürig, J., Bruhn, T., Zurborn, K-H, Gutensohn, K., Bruhn, H.D. and Béress, L. (1997), *Thromb. Res.*, 85:6. 479-491.

[93] Mulloy, B., Ribeiro, A.-C., Alves, A.-P., Vieira, R.P. and Mourao, P.A.S. (1994), *J.Biol.Chem.*, 269. 22113-22123.

[94] Mourao, P.A.S., Pereira, M.S., Pavao, M.S.G., Mulloy, B., Tollefsen, D.M., Mowinckel, M-C. and Albigaard, U (1996), *J.Biol.Chem.*, 271:39. 23973-23984.

[95] Pavao, M.S.G., Mourao, P.A.S., Mulloy, B. and Tollefsen, D.M. (1995), *J.Biol.Chem*, 270. 31027-31036.

[96] O'Neill, A.N. . (1954), *J.Am.Chem.Soc.*, 76. 5074-507.

[97] Nardella, A., Chaubet, F., Boisson-Vidal, C., Blondin, C., Durand, P. and Jozefonvicz, J. (1996), *Carbohydr.Res.*, 289. 201-208.

[98] Mabeau, S., Kloareg, B. and Joseleau, J-P(1990), *Phytochemistry*, 29. 2441-2445.

[99] Nardella, A., Chaubet, F., Sinquin, C., Colliec-Jouault, S., Boisson-Vidal, C., Durand, P. and Jozefonvicz, J. (1996).French Patent 95.10045 (1995), Int.Patent PCT/FR/96/01314

[100] Medcalf, D.G. and Larsen, B (1977), *Carbohydr.Res.*, 59. 531-537.

[101] Nagasawa, K, Uchiyama, H., Sato, N, and Hatano, A (1992), *Carbohydr.Res.*, 236. 165-180.

[102] Volpi, N., Mascellani, G. and Bianchini, P (1992), *Anal.Biochem.*, 200. 100-107.

[103] Fenton, H.J.H. . (1894), *J.Chem.Soc.*, 65.899-910.

[104] Uchiyama, H., Dobashi, Y., Ohkouchi, K. and Nagasawa, K. (1990), *J.Biol.Chem.*, 265. 7753-7759.

[105] Mulloy, B., Forster, M.J., Jones, C. and Davies, D.B. (1993), *Biochem.J.*, 293. 849-858.

[106] Lam, L.H., Silbert, J.E. and Rosenberg, R.D(1976), *Biochem.Biophys.Res.Commun.*, 69. 570-577.

[107] Sinniger, V., Tapon-Bretaudiere, J., Millien, C., Muller, D., Jozefonvicz, J. and Fischer, A-M. (1993), *J.Chromatogr. Biomed. Appl.*, 615. 215-223.

[108] Cabassi, F., Casu, B. and Perlin, A.S. (1978), *Carbohydrate Research*, 63. 1-12.

[109] Anderson, N.S., Dolan, T.C.S., Penman, A., Rees, D.A., Mueller, G.P., Stancioff, D.J. and Stanley, N.F. (1968), *J.Chem.Soc., C.* 602-606.

[110] Mori, H. and Nisizawa, K. (1982), *Bull. Jpn. Soc. Sci. Fish.*, 48:7. 981-986.

[111] Cael, J.J., Isaac, D.H.Blackwell, J., Koenig, J.L., Atkins, E.D.T. and Sheehan, J.K. *Carbohydr. Res.*, 50 (1976), 169-179.

[112] Ribeiro, A.-C., Vieira, R.P., Mourao, P.A.S., and Mulloy, B. (1994), *Carbohydr.Res.*, 255. 225-240.

[113] Pavao, M.S.G. and Mourao, P.A.S. *Carbohydr.Res.*, 208 (1990), 153-161.

[114] Percival, E. (1979), *Br.Phycol.J.*, 14 .103-117.

# Chapter 7

# Immune stimulating properties of di-equatorially $\beta(1\rightarrow4)$ linked poly-uronides.

G.SKJÅK-BRÆK, T.FLO, Ø.HALAAS and T.ESPEVIK
*Institute of Biotechnology and Institute of Cancer Research and Molecular Biology, Norwegian University of Science and Technology, N-7005 Trondheim, Norway*

Key words: alginate, cytokines, immunestimulation,

Abstract

The biological activities of complex carbohydrates and polysaccharides have traditionally been attributed to short oligosaccharide structures. In the last decade several reports have been publishes suggesting that biological activity, i.e antitumour activity as well as the adjuvance effect of polysaccharides of various structures and origins is depending upon certain macromolecular structures. The best known example is the β-1-3-linked glucan. We have previously found that certain alginates induce human monocytes to produce TNF, IL-1 and IL-6, and that the cytokine inducing ability depends on the mannuronic acid (M) content as well as the molecular weight of the alginate. Our data demonstrate that alginate enriched in mannuronic acid were the cytokine inducing polysaccharides whereas guluronic acid residues did not stimulate monocytes to produce cytokines. Similar effects are found for other polyuronides containing β-1-4 di-equatorial linked sequences. High M-alginate and lipopolysaccharide (LPS) were found to stimulate human monocytes by similar mechanism, which involved the CD14 LPS/LBP receptor. The mechanism for the interaction between the polyuronides and the cytokine producing cells will be discussed. Defined polysaccharides, which specifically stimulate the non-specific immune system, may be important agents for treatment of various infectious diseases. The potent cytokine inducing ability of ß1-4 linked uronic acid polymers on monocytes *in vitro* implicates possible interesting effects *in vivo*. The effect of high M alginate and C-6 oxidised cellulose in various *in vivo* models, ranging from bacterial sepsis in rodents to adjuvance effects in marine fishes have been tested.



*B.S. Paulsen (ed.), Bioactive Carbohydrate Polymers,* 85–93.

## 1. Cytokine stimulating properties of β 1->4 linked uronic acid polymers

The polyuronide structures discussed in this paper are found exclusively in various amounts in a group of biopolymers designated alginates. Alginates are glycuronans extracted from seaweed or produced by some bacteria. The molecules are linear chains of (1 - 4)-linked residues of β-D-mannuronic acid and α-L-guluronic acid in different proportion and sequential arrangements. The most common arrangement is that of a block copolymer, in which long, homo-polymeric sequences of M-residues ("MM-blocks") and similar sequences of G-residues ("GG"-blocks) are interspaced between sequences of mixed composition ("MG-blocks"). See Figure 1.

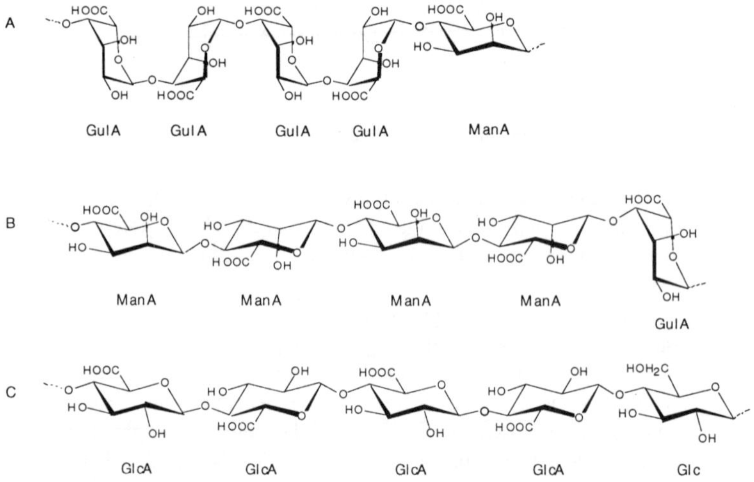


*Figure 1.* Structure of polyuronides: A) G-blocks, B. M-blocks and C) C6OXY

They form gels with cations and the gel-forming capacity correlates with the content and length of the G-blocks. Entrapment of cells and tissues within spheres of calcium alginate gels is a widely used technique for immobilisation and subsequent transplantation. Microencapsulation of Langerhans islets has been used to treat diabetes in animal models [1] and now also in humans [2]. One of the main problems with injection of alginate capsules is overgrowth of the capsules with phagocytes and fibroblasts, which resembles a foreign body/inflammatory reaction. This reaction has been attributed to an immune response to cellular material leaking out through the capsule membrane or exposure through capsule breakage. Since fibrosis could be observed also for empty capsules [3] we searched for immune stimulating compounds.

## *Correlation between polymer structure and immune stimulation*

When alginates were tested for cytokine induction on human monocytes it became evident that the ability of alginates to induce TNF, IL-1 and IL-6 correlated with the M-content of the alginate, as well as on the molecular size [4]. This is illustrated in Figure. 2a, where alginates with different contents of mannuronic acid were tested for induction of TNF from human monocytes. The highest potency is found for polymers containing more than 95 % mannuronic acid residues and with a molecular size above 50 000 Dalton. This material were isolated from *Pseudomonas aeruginosa* and is designated poly-M. An analog structure, a di-equatorially β-1-4 linked D-glucuronic acid, prepared by selective oxidation at C-6 in cellulose, also stimulate monocytes to produce TNF although with less potency than poly-M [4 ].

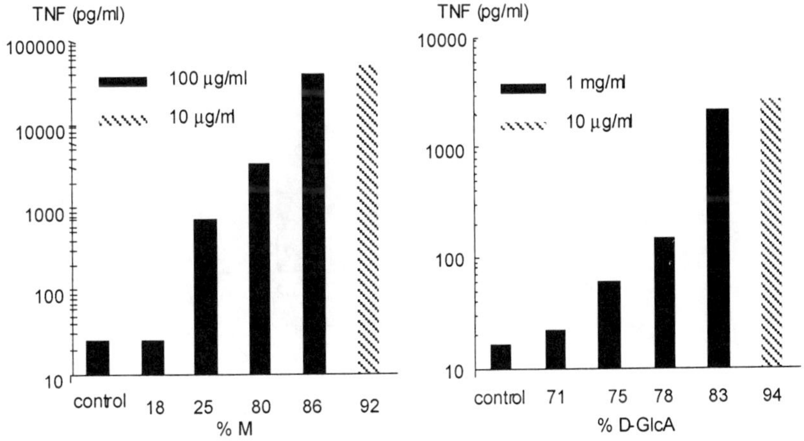


*Figure 2.* Stimulation of TNF production .A) Effect of the M-content in alginate, B) Effect of the GlcA content in C-6OXY

The 3D structure of C6OXY (94 % D-GlcA) is similar to that of poly-M except that the consecutive uronic acid residues in C6OXY are broken up with D-Glc residues. The stimulatory effect of C6OXY on TNF production depends on the amount of D-GlcA in the, suggesting that the TNF induction from monocytes may occur with different types of ß1-4 linked uronic polymers.

Increase in mRNA for other cytokines like M-CSF, GM-CSF and IL12 p40 are also observed in poly-M treated monocytes[7]. Furthermore, treatment of high M alginate with a recombinant mannuronan C-5-epimerase AlgE2 which, converts mannuronic acid sequences into G-blocks resulted in loss of the TNF inducing ability. This result also strengthen the conclusion that mannuronic acid is the active cytokine inducing residues in alginate.

The TNF-inducing capacity also increases with the molecular weight of the polymer up to 50000 Dalton. By degrading the polymer to oligomers with a $DP_n \approx 20$ either with a controlled acid hydrolysis or by treatment with a M-specific lyase the TNF inducing capacity is lost. Immune stimulation can however be regained and even strongly potentiated by linking the oligouronides covalently to microparticles as shown in Figure 3 [ 5]

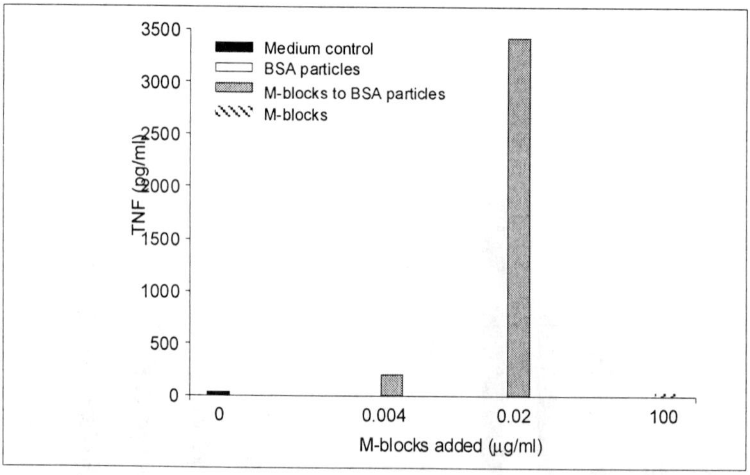


*Figure 3* Effect of M-blocks linked to BSA -particles on the TNF production in human monocytes

## *Effects of uronic acid polymers in different in vivo models.*

Defined polysaccharides, which specifically stimulate the non-specific immune system, may be important agents for treatment of various infectious diseases. The potent cytokine inducing ability of ß1-4 linked uronic acid polymers on monocytes *in vitro* implicates possible interesting effects in different *in vivo* models. Little has been reported on immune stimulating effects of alginates in animal models. The observation that alginates enriched in mannuronic acid, but not guluronic acids, are the active cytokine inducing residues in alginate extends the findings by Iizima-Mizui et al. [6] who found that the most active polysaccharide fraction for antitumor activity was alginate with a high mannuronic acid content. Table 2 summarizes our most important data on the biological effect of poly-M and C6OXY. Injection of C6OXY in the peritoneal cavity results in a transient production of both TNF and IL-6, which peaks 60 min after the injection (Shalaby, R., Skjåk-Bræk, G. and Espevik, T., unpublished data). These data demonstrate that ß-1-4 linked uronic acid polymers also are active cytokine inducers *in vivo*. Of particular interest is that poly-M can protect mice against lethal infection with *E. coli* or *S. aureus* (Espevik, T. and Skjåk-Bræk, G, unpublished data) . In line with these observations, poly-M gives a marked protection of mice against lethal irradiation [7]. Exposing animals to irradiation leads to loss in the ability of the bone marrow to generate white blood cells. Thus, irradiated animals may die from infections caused by bacteria, which normally are well tolerated. See Figure 4. In combination with sub-optimal concentration of colony stimulating factors, poly-M enhances the formation of GM-CSF colonies suggesting that poly-M can increase the production of myeloid blood cell [7]. This could be one mechanism behind the radioprotective effect of poly-M. The stimulating effect of poly-M on hematopoietic cells may be clinically important.

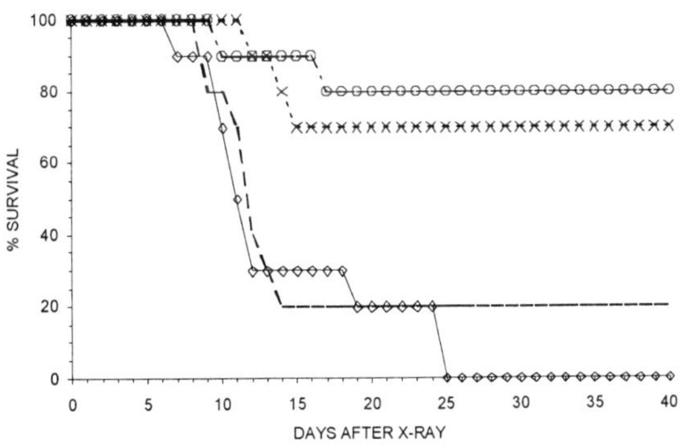


*Figure 4.* Effect of prophylactic (-24h, intraperitoneal ) administration of mannuronan on survival of lethally irradiated (7.3Gy) C57B1/6 mice, 10 mice/group. Mannuronan was administrated in 0.5ml 0.9%NaCl.](-◇-) NaCl, (-□-) 0.5mg/kgbw), (-O-) 1mg/kgbw), (-x- ) 2mg/kgbw)

Alginate rich in mannuronic acid also has immune stimulating properties in fish. Juvenile turbots (*Scophthalmus maximus* L.) [8] fed with alginate rich in mannuronic acid have increased protection against one type of bacteria which are pathogenic for turbots.

From these data, it is evident that ß-1-4 linked uronic acid polymers have potent biological effects in several biological systems, and that these effects probably are caused by stimulation of the monocytes/macrophages.

## *Mechanisms of poly-M effects - comparison with LPS*

Since poly-M stimulates monocytes in a way that resembles lipopolysaccharide (LPS) from gram negative bacteria, it was important to compare mechanisms of poly-M and LPS effects..

Responses of both LPS and poly-M involve CD14 on the monocyte membrane [3]. Poly-M binds to CD14 which also expresses a binding site for one to two LPS molecules [9]. We have previously reported that poly-M binds to CD14 on monocytes and that this binding occurs in the presence of serum [9]. The binding of poly-M and LPS to monocytes can be inhibited by addition of G-blocks, which also suggests a common binding site for these apparently different polysaccharides [4]. After our initial observation several reports have now implicated a role for CD14 in responses to a variety of different compounds such as soluble peptidoglycan fragments and protein free phenol extracts from *S. aureus* [10,11], rhamnose-glucose polymers from *Streptococcus mutans* [12], chitosan from arthropods [13] , and insoluble cell walls from different gram positive bacteria [14]. These data implicate that CD14 has a broad specificity for compounds that are rich in different types of sugar residues. Thus, CD14 has been implicated as a basis recognition molecule for innate immunity [14].

Lipopolysaccharide can also stimulate cells that do not express membrane CD14, which indicates that a wide variety of different cell types are affected by LPS. For example, LPS activates endothelial cells and also U373 astrocytoma cells which do not express membrane CD14. We found that CD14 is essential for response of each of these cell types to LPS, but that the CD14 is derived from soluble CD14, which is present in μg quantities in serum, and not from membrane CD14 [38]. In contrast to LPS, poly-M is not able to stimulate cell types, which lack membrane CD14 such as U373, cells [9]. These data implies that LPS interacts with cells in several ways, which may explain why LPS affects different cell types. This broad stimulatory pattern of LPS is likely to be important for its lethal effect *in vivo.*

The apparent specific effect of uronic acid polymers on CD14-positive cells may implicate low systemic toxicity and suggests potentially interesting applications of as an immunomodulator. Thus, poly-M may activate the non-specific immune system resulting in increased protection against various types of infections. The immunological effect of polyuronic acid and the mechanism of action are summarised in Table 1 and Table 2.

## Table 1. Effects of poly uronic acids on immune cell functions -biological effects

- Induction of TNF, IL-1, IL-6, GM-CSF, and IL-12 p40 in human monocytes
- Induction of TNF and IL-6 in mice
- Protection against lethal gram positive and gram negative infections
- Protection against lethal effects of irradiation
- Increases the generation of myeloid progenitor cells
- Increases the amount of antibody producing cells
- Increases non-specific immunity in turbots

## Table 2. Stimulation of cells with poly-M - mechanisms of action

- Poly-M binds to CD14 on monocytes in the presence of serum
- Antibodies against CD14 inhibits poly-M induced TNF production
- Poly-M stimulates monocytes, but not LPS responsive cells which lack membrane CD14
- Poly-M binds to LBP and BPI[15]
- Induction of TNF from monocytes by poly-M is enhanced by LBP
- LBP enhances binding of poly-M to monocytes
- BPI inhibits poly-M induced TNF production from monocytes
- BPI inhibits binding of poly-M to monocytes.

## 3. Sources of polymers with a high content of β 1-4 linked mannuronic acid.

Immune stimulating polyuronides can be isolated from both algal and bacterial sources

*Algae.*

Even though the primary product in the algal biosynthesis is believed to be homopolymeric mannuronans, the material extracted from brown algae invariably contains guluronic acid both in blocks and as single G residues.. The highest M containing polymer (85% M) is found in the fruiting bodies of *Ascophyllum nodosum* and *Fucus vesiculosus*. This material is an immune stimulant but much less potent than the bacteria materials with more than 90 % and only singles G residues.

*Bacteria.*

Alginates or alginate-like polymers, with more extreme compositional features are produced as exocellular material by several bacteria. Mucoid mutants of *Pseudomonas aeruginosa* isolated from the lung of patients suffering from cystic fibrosis produces exopolymers that resembles alginates in being a co-polymer of 1-4 linked D-mannuronic and L-guluronic acid. The bacterial polymer is substitutetd with O-acetyl groups and the lacks consecutive G-residues (G-blocks). (The O-acetyl groups are easily split off by treatment with alkali, a treatment that also deactivates LPS). Polysaccharide production is however, not common in P. *aeruginosa* outside the special environment of the lungs of infected patients and mucoid colonies seem to revert to non-mucoid when grown on liquid media in culture. Recently great advances have been made in the genetics of alginate biosynthesis in *P. aeruginosa*, by identifying most of the genes coding for the alginate biosynthesis. Including AlgG, which encodes a Ca-independent mannuronan C-5 epimerase. We are currently working with AlgG negative mutants from P. aeruginosa as well as from other non-pathogenic psedomonads that produce a homopolymeric mannuronan.

A similar, partly acetylated, polyuronide is produced by the soil bacterium *Azotobacter vinelandii*. These latter polymers is a true block co-polymers of M and G, containing homo-polymeric regions of M and G interspaced with regions containing both types of monomers.

Since this bacterium encodes eight different C-5 epimerases, negative mutants are difficult to isolate.

Some compositional data of various types of alginates are summarised in Table 3.

Table 3

| **Source** | **Composition** $F_M$ | **Sequence** $F_{GG}$ | **AlgG÷mutant** |
|---|---|---|---|
| Seaweed | 0.20-0.85 | 0.07-0.70 | |
| *A. vinelandii* | 0.15-0.85 | 0.02-0.80 | |
| *P. aeruginosa* | 0.5-1.0 | 0 | Yes |
| *P. mendocina* | >0.5 | 0 | |
| *P. putida* | >0.5 | 0 | |
| *P. fluorescens* | 0.5-1.0 | 0 | Yes |
| *P.corrugata* | >0.5 | 0 | |
| | | | |

## References


[1] P. Soon-Shiong, E. Feldman, R. Nelson, R. Heintz, Z. Yao, Q. Yao, T. Zheng, N. Merideth, G. Skjak-Braek, T. Espevik, O. Smidsrød and P. Sandford, (1993) *PNAS* 90. 5843-5847.

[2] P. Soon-Shiong, R.E. Heinz, N. Merideth, Q.X. Yao, Z. Yao, T. Zheng, M. Murphy, M.K. Moloney, M. Schmehl, M. Harris et al., (1994) *Lancet* 343. 950-951.

[3] M. Otterlei, K. Østgaard, G. Skjak-Braek, O. Smidsrød, P. Soon-Shiong and T. Espevik, (1991) *J. Immunother.* 10. 286-291.

[4] M. Otterlei, A. Sundan, G. Skjåk-Bræk, L. Ryan, O. Smidsrød and T. Espevik. (1993) *Infect. Immun.* 61. 1917-1925.

[5] G. Berntsen, L. Kilaas, T.H. Flo, A. Medvedev, G.Skjåk-Bræk A. Sundan and T. Espevik (1998) *J.Clinical and Diagnostic Laboratory Immunology*, 5. 55-361

[6] N. Iisima-Mizui, M. Fujihara, J. Komiyama, I. Umezawa and T. Nagumo, (1985) *Kitasato Arch. Exp. Med.* 58. 59

[7] Ø. Halaas, W.M. Olsen, O.P. Veiby, D. Løvhaug, G.Skjåk-Bræk and T. Espevik (1997) *Scand. J. of Immunol*, 46. 358-365

[8] J. Skjermo, T. Defoort, M. Delasque, T. Espevik, Y. Olsen and G. Skjak-Braek, (1995) *Fish & Shellfish Immunol.* 5. 531-534.

[9] T. Espevik, M. Otterlei, G. Skjåk-Bræk, L. Ryan, S.D. Wrigth and A. Sundan, (1993) *Eur. J. Immunology* 23. 255-261

[10]. B. Weidemann, H. Brade, E.T. Rietschel, R. Dziarski, V. Bazil, S. Kusomoto, H.D. Flad and A.J. Ulmer, (1994) *Infect. Immun.* 62. 4709-4715.

[11] T. Kusunoki, E. hailman, T.S.-C. Juan, H.S. Lichenstein and S.D. Wright, (1995) *J. Exp. Med.* 182. 1673-1682.

[12] M. Soell, E. Lett, F. Holveck, M. Scholler, D. Wachsmann and J. Klein, (1995) *J. Immunol.* 154. 851-860.

[13]. M. Otterlei, K.M. Vårum, L. Ryan and T. Espevik, (1994) *Vaccine* 12. 825-832.

[14]. J. Pugin, D. Haumann, A. Tomasz, V.V. Kravchenko, M.P. Glauser, P.S. Tobias and R.J. Ulevich, (1994) *Immunity* 1. 509-516.

[15] T.G. Jahr, L. Ryan, H.S. Lichtenstien, G. Skjåk-Bræk and T. Espevik, (1997) *Infection and Immunity* 65. 89-94

Chapter 8

# Immunostimulatory ß(1,3)-D-glucans; prophylactic drugs against threatening infectious diseases of fish

R: DALMO
*Institute of Marine Biochemistry, Norwegian College of Fishery Science, University of Tromsø, N-9037 Tromsø, Norway*

Key words: ß(1,3)-D-glucans, immunostimulants, aquaculture

Abstract: Homopolysaccharides, consisting of one sugar, are described by use of the suffix "an". Thus a polyglucose (polyglucopyranose) is termed "glucan"; others are heteropolysaccharides that may contain several different constituent residues. ß(1,3)-D-Glucans (GLs) are the most commonly used term for homopolysaccharides that has ß(1,3)-D-linkages in the backbone, and may possess ß-D-glucosidic linkages at position 6 in different, often repeating units (branches). In the nature GLs are widespread and are there found in plants, micro-algae, bacteria, yeast and mushrooms. GLs vary in length and their degree of branching. Curdlan (*Alcaligenes faecalis* var. *myxogenes* IFO 13140) is a linear, unbranched GL, whereas scleroglucan and SSG (*Sclerotium glucanium* and *Sclerotinia sclerotiorum* IFO 9395, respectively), schizophyllan (*Schizophyllum commune*) and lentinan (*Lentinus edodes*) have a main chain composed of ß(1,3)-D-glucopyranosyl residues with residues carrying a ß(1,6)-D-linked single glucopyranosyl group. Laminarin (*Laminaria digitata*) and laminaran (*Laminaria hyperborea*) have more or less extensive branches through ß(1,6)-D-linkages from the backbone. Yeast glucan is a particulate carbohydrate that consists of glucose and mannose. GL's aqueous solubility ranges from insoluble (curdlan at room temperature) via gelling (scleroglucan) to fully soluble (laminarin and partly laminaran). The solubility in water is dependent on temperature [1, 2]. The use of GLs as immunostimulants in aquaculture is discussed.

## 1. Introduction

An immunomodulator/immunostimulant may be defined as a substance that stimulates leucocytes- particularly cells of the macrophage system and thereby modulates, most often potentiates, the body's immune system [3]. The term immunomodulator is often used interchangeably with



*B.S. Paulsen (ed.), Bioactive Carbohydrate Polymers,* 95–106.

immunostimulants, adjuvants and biological response modifier. An immumodulator may be given alone in order to alter non-specific defence mechanisms, or it may be administered together with another substance (e.g vaccine component) to alter both the non-specific defence mechanisms and the specific immune response. The non-specific defence mechanisms are important in the protection of all multicellular animals against invading pathogenic microorganisms. Phagocytes play a central role in the non-specific defence and are present throughout the animal kingdom ranging from wandering leucocytes in acoelomates to some coelomocytes or hemocytes in invertebrates and to mononuclear phagocytes in vertebrates [4, 5]. Most immunmodulating compounds are of fungal, bacterial or plant origin, such as lipopolysaccharide, peptidoglycans, muramyl dipeptides and GLs [3, 6, 7]. When immunostimulants are administered to an animal, the defence system may act in the same way as if it was challenged with living micro-organsims, and thus initiate reactions necessary to eradicate the aetiologic agent. However, the biological activities of the immunstimulants are so multiple and potent that some of them are more harmful than beneficial. Negative side-effects include pyrogenicity [8], induction of polyarthritis [9] and induction of granulomateous reaction [10].

In 1969 Chihara and coworkers reported that a systemic injection of lentinan led to arrest of tumor growth in transplanted mouse tumors [11]. Since then, there have been a great number of reports of similar effects with other polysaccharides.

In numerous studies the relationship between the chemical structure and biological activity has been addressed. In previous studies it was assumed that GLs that possessed triple helical conformation were more biologically active than the corresponding low molecular sized GLs which often possess more random conformation in water [1]. The helical structure is dependent on the molecular mass and the degree of branching. However, in a recent study no correlation between the degree of helical structure and the biological activity was found [2], and it is supposed that the molecular mass and ratio of branching points (glucose residues) is more important than their conformation to augment biological activity [12].

In conclusion, the biological activity, measured as the profile of cytokine production, production of other proinflammatory substances, anti-cancer and anti-infectious effects, *in vivo* or *in vitro*, of GLs that are to be investigated is dependent on chemical structure and conformation of GL.

It is well established that macrophages are central cells in the immune system, as e.g antigen presenting cells initiating the antibody response and as multifunctional cells that have capability to secrete a great number of products that have different biological effects. A significant increase in macrophage property that is related to an effector mechanism (e.g bactericidal) is termed macrophage "activation" [13, 14]. Macrophages are often target cells for immunostimulants such as GLs, and the effect GLs have on the cells may results in macrophage activation. Upon activation, antimicrobial defence mechanims are in alerted position waiting for pathogens that may be come.

## 2. GL receptors

The biological activity of GLs is thought to be dependent on a GL specific receptor [15, 16] on macrophages, monocytes, neutrophilic granulocytes, minor subsets of B and T cells, and natural killer (NK) cells. The GL binding protein is reported to be lectin site(s) located C-terminal to the CD11b I-domain of the complement receptor 3 (CD11b/CD18; Mac-1; $\alpha_m$ $\beta_2$-integrin). Its sugar specificity is broader than originally believed, allowing it to react with certain polysaccharides containing mannose or N-acetyl-D-glucosamine, as well as glucose [17, 18]. After binding, certain carbohydrates such as GLs or lipopolysaccharides (LPS) can bind to and activate CR3, allowing the receptor to assume its activated state that may result in killing/destruction of C3i (the third complement breakdown product) bearing particles/cells [19]. The CR3-like receptors that bind zymosan, (a particle that consists partly of GL and mannan) are also described on channel catfish (*Ictalurus punctatus*) neutrophilic granulocytes. No binding of the fluorescence labelled zymosan occured when the neutrophils were cultivated without serum, with heat inactivated serum (complement activation is not functioning), and with GL [20]. Furthermore, macrophages obtained from Atlantic salmon (*Salmo salar* L.) have also been shown to carry GL receptors, and the authors speculated about the presence of both GL receptor and CR3 [21].

After GL receptor binding on mammalian leucocytes, signal transduction may occur (via protein kinase C, tyrosine kinases, phospholipase $A_2$) [22-26], and a variety of transcription factor activation may occur which eventually would result in protein synthesis. Many of the newly synthetised are proinflammatory substances that have pleiotropic effects on the host´s response to e.g infectious microbes. Thus, GL may induce the production of benificial or harmful substances from the target cells.

In mammals, numerous studies have elucidated the modulating effects of GLs on the hosts defence system against invading pathogens. Several GLs and chemical derivatives (e.g sulfated, phosphorylated, carboxymethylated, hydroxyethylated) of them have shown anti-infectious effects (bacteria, virus and parasites), especially when used as prophylactic drugs [1].

## 3. *In vitro* effects of GLs

From the mid of 1980´s, the potential GLs have as immunostimulants or immunomodulators in fish rearing system has been evaluated. The GL that most frequently has been tested for its *in vitro* effects on fish leucocytes is MacroGard®. This polysaccharide, obtained from *Saccharomyces cerevisiae*, has a ß1,3-linked backbone with ß1,3- and ß1,6-linked side chains [21]. The microparticulate (3-4 µm) GL has been reported to increase the level of respiratory burst activity by phorbol myristate acetate (PMA) or zymosan elicited leucocytes (macrophages and neutrophils). The macrophages showed a marked increase in the respiratory burst activity 4 to 7 d after addition of 1 µg $ml^{-1}$ GL. However, after the cells were incubated with higher amounts of the particulate (10 µg $ml^{-1}$), no such effect could be observed. A high MacroGard® concentration in the cultures led to a inhibitory effect measeured as production of superoxide anion ($O_2^-$) [27]. The MacroGard® treated cells did not show increased bactericidal activity against the avirulent and virulent strains of *A. salmonicida*. It was therefore concluded that the respiratory burst activity should not be considered sufficient in enhancing the bactericidal activity. Other factors may be needed to kill *A. salmonicida* [27]. The MacroGard® has been shown to enhance the respiratory burst activity also in dab (*Limanda limanda*) leucocytes [28]. Furthermore, after exposure to VitaStim® *in vitro*, elicited blood and anterior kidney derived neutrophils showed increased respiratory burst activity compared to untreated cells [29]. Relatively few studies have been performed to elucidate the effect soluble glucans have on fish leucocytes *in vitro*. In one such study, curdlan that has been formolysed (hydrolysed) and subsequently aminated (AG) has been reported to stimulate Atlantic salmon macrophages *in vitro*. Activation of these cells was judged by their respiratory burst response to AG, increased uptake of a pinocytic marker (neutral red), and the activity of lysosomal acid phosphatase was also elevated compared to control cells [30]. The molecular weight of AG has been approximated to be less than 25 KDa [31]. The stimulatory effect of laminaran, obtained from *Laminaria hyperborea*, on Atlantic salmon macrophages has been presented [32]. As the former GLs (MacroGard® and AG), an addition of laminaran to the cells resulted in higher respiratory burst activity compared to untreated cells. The activity of the lysosomal enzyme acid phosphatase and the rate of cell "drinking" (non- receptor mediated pinocytosis), considered as markers for cell activation, were also hightened compared to control cells. A sulfated derivative of laminaran was also tested for its ability to stimulate the macrophages, but possessed lesser biological activity than the activity of laminaran. The molecular weight of the preparation of laminaran used in these studies was approximately 4000, and was reported to be water soluble to some extent. In contrast to Jørgensen &

Robertsen (1995) [27] who reported that a concentration of 1 µg $ml^{-1}$ in the cell cultures was the most efficient dose, the corresponding optimum doses of laminaran and AG were found to be 20µg $ml^{-1}$. It seems that MacroGard® per dry weight basis is more efficient to stimulate macrophages *in vitro*, but it should be emphasised that it is difficult to compare the biological activity possessed by particulates and water soluble GLs.

## *4. In vivo* effects of GLs

In general, the GLs that have been tested for their *in vivo* activity are injected intraperitoneally to fish, and som GLs are immersed in water (given to fish via the drinking water). The injection into the peritoneal cavity of fish is a relative simple method of antigen delivery, but handling stress often results is some losses. Immersion is more easy to perform, but demand higher antigen dose compared with the amount of injected vaccines. Peroral administration of antigen via feed is the most desireable method of vaccination, but many studies have pointed out the lack of effects concerning survival after bacterial infections. This mode of antigen delivery demands higher amounts of antigen compared to the formers.

Many studies concerning the *in vivo* effects of different GLs have been performed. In most cases fish have been injected i.p with the immunstimulator and the effect of the GLs have been evaluated as modulation of the activity to the cells of interest. Insoluble/particulate GLs (MacroGard®, VitaStim® and barley ß-glucan) have shown promising effects as immunostimulants to increase the antibactericidal activity of leucocytes. Intraperitoneally (i.p) injected MacroGard® resulted in increased bactericidal activity by macrophages extracted from the anterior kidney (the major hematopoietic organ in fish) of Atlantic salmon [33]. Also enhanced bacterial killing by macrophages was observed after this substance was given i.p to rainbow trout [34]. A study, employing MacroGard®, showed that rainbow trout macrophages phagocytosed higher number of *Vibrio salmonicida* after the fish were i.p injected by MacroGard® [35]. However, the level of phagocytosis of another fish pathogen, *Renibacterium salmoninarum* was unchanged. The fact that one of these two pathogens was phagocytosed in higher number, indicate that some receptors that mediate phagocytosis were upregulated. In another fish species, channel catfish, baker´s yeast was i.p injected. After the treatment, the phagocytic leucocytes displayed increased phagocytic and bactericidal activity [36].

The phagocytic activity by blood leucocytes, and by phagocytic cells in organs that constitute the reticuloendothelial system (RES) (macrophages) were apparently stimulated when schizophyllan, scleroglucan and lentinan were i.p injected to carp (*Cyprinus carpio*) as reported by [37]. A more rapid blood clearance of bacteria that were injected was observed. A combination of O-antigen (probably LPS enriched bacterin) and yeast glucan that were injected i.p, was also efficient in increasing the bactericidal activity by turbot (*Scophthalmus maximus*) leucocytes [38]. In a study by Chen & Ainsworth (1992) [36], no differences between the amount of hydrogen peroxide produced by cells obtained from baker's yeast treated channel catfish compared with control cells were observed. In contrast, rainbow trout phagocytes produced more hydrogen peroxide after an i.p injection of MacroGard® than cells obtained from control fish [35].

A common feature of GLs is that they induce an alternative complement cascade activation in fish [37, 38] after parenteral injections. However, after giving the GLs perorally no such complement activation has been observed [39-41]. Also the lysozyme levels in blood plamsa/serum is affected by parenteral GL treatment. In many studies performed in different fish species, the lysozyme levels have been found to increase after the treatment [33, 35, 38, 42, 43]. The lysozyme levels after peroral administration of GLs have been reported to be unchanged in rainbow trout (yeast glucan) [40, 41] and in dentex (*Dentex dentex*) (MacroGard® and VitaStim®) [44], and to be increased in turbot [MacroGard®+Vibriffa bain® (a commercial anti-vibriosis vaccine from Rhone-Merieux, Lyon, France)] [39] and the lysozyme levels were observed to be increased, although not significant, in channel catfish [29]. Other components of the non-specific defence mechanisms that have been reported to increase in their activity after peroral administration of GLs are: respiratory burst activity, myeloperoxidase- and phagocytic activity [45], blood leucocyte number [39], chemiluminescence response by phagocytes and antibody formation [40], phagocyte migration [46], respiratory burst activity and lysosomal acid phosphatase activity [47]; adjuvant activity on rainbow trout antibody response after vaccination with *Yersinia ruckeri* [41] and enhanced effect on concanavalin A induced prolifaration of rainbow trout lymphocytes [41]. In contrast to these findings, the chemiluminescence response of phagocytes obtained from turbot and was unchanged compared to control cells after peroral administration of yeast glucan, and yeast glucan together with a vibriosis vaccine preparation [39]. The same conclusion was also drawn after rainbow trout were fed on MacroGard®, where anterior kidney phagocytes did not show increased respiratory burst activity [41]. In the same study, blood lysozyme levels and the pinocytic activity of phagocytes obtained from fish fed MacroGard® were unchanged [41]. Blood leucocyte number, respiratory burst activity, and plasma protein levels were not significantly increased compared to controls after dentex were treated with yeast glucan or VitaStim® perorally [44]. It seems reasonable to conclude that some inter-species differences, with respect to the efficiency of GLs to elevate non-specific defence mechanisms, must exist. The dose, duration of treatments, stress and sensivity of the assay systems should be important factors to find out whether the GLs have any effect or not. A few experiments have been performed in evaluating the effects the GLs have, after peroral administration, on the organsims non-specific defence mechanisms. Such effects may result in an increase of the survival rate after experimental infections. In two of these studies, no increased survival after a *Vibrio anguillarum* infection was observed after turbots were p.o administered with yeast glucan [39], and the same conclusion was drawn after channel catfish were treated with yeast glucan or barley ß-glucan perorally, and subsequently challenged with *Edwadsiella ictaluri* [46]. Furthermore, Atlantic salmon fed on laminaran enriched diet did not show increased survivial rate after experimental challenge with *Vibrio salmonicida* and *A. salmonicida* infections [48]. In contrast to these reports, Nikl *et al.* (1993) observed that *Oncorhynchus tsawytschka* increased their resistance towards an experimental *A. salmonicida* infection after the fish were fed on VitaStim® enriched diet [49]. Finally, Siwicki *et al.* (1994) also showed that perorally administration of yeast glucan to rainbow trout led to increased survival after an *A. salmonicida* infection [45].

Interestingly, an anti-protozoic effect was displayed by fish (dentex) perorally treated with yeast glucan and VitaStim® [44].

Furthermore, rainbow trout that were treated with a MacroGard® injection followed with an injection of convalescent sera from fish that have survived IHNV infection (infectious hematopoietic necrosis virus) resulted in increased protection after IHNV challenge compared to control fish (fish injected only with convalescent sera or saline) [50].

Based on these results from the experiments employing GL as feed additive, it is difficult to make any conclusion regarding the effects GLs have when the purpose is to increase the survival rate after fish have been perorally treated with the immunostimulants.

It is evident that i.p injected GLs are far more effective to increase the non-specific defence mechanisms and thus increasing the survival rate compared to GLs that are perorally administered. In this context, carp injected with schizophyllan, scleroglucan or lentinan resisted *E. tarda* and *A. hydrohila* infections [37, 38]. The polysaccharides schizophyllan and scleroglucan were also observed efficient in elevating non-specific defence mechanisms leading to increased survivial of Yellowtail after an experimental *streptococcus* sp. infection. No such positive effect was seen after *Pasteurella piscicida* infection [42]. Barley ß-glucan has also shown to increase survival of tilapia and brook trout (*Salvelinus fontinalis*) that were treated by an i.p injection and subsequent challenged with *A. hydrophila, E. tarda* and *A. salmonicida* infections, respectively [51, 52]. Intraperitoneal injection of MacroGard® in Atlantic salmon resulted in enhanced survivial rate after *Yersinia ruckeri*, *V. anguillarum* and *V. salmonicida* infections [53]. In a study performed by Nikl and coworkers, seven immunostimulants that were i.p injected to coho salmon were evaluated for their efficacy in potentiating a vaccine preparation (formalin-killed bacteria). Among the immunostimulants, VitaStim® and lentinan were shown to significantly increase the survival rate of the fish compared to fish given the vaccine alone [54]. Also laminaran has showed promising effects on blue gourmami (*Trichogaster trichopterus*) survival after the fish were i.p injected and subsequently infected with *A. hydrophila* infection [55].

In conclusion, many GLs have been reported to possess immunstimulatory activites, and to date, and among the GLs tested, only curdlan and laminarin (called laminaran in the reference) have been found not to increase the survival rate to fish after i.p injection with these substances [38].

## 5. Tissue- and organ distribution of GLs

The intestinal uptake and tissue distribution of aminated curdlan (AG) after peroral and peranal administrations to Atlantic salmon have been described by Ingebrigtsen *et al.* (1993) and by Sveinbjørnsson e*t al.* (1995) [56,57]. By the employed methods of antigen delivery (AG administered in a cellulose gel), AG was confined only in the intestinal tissue after peroral administration [56], but liquid scintillation counting experiments revealed that AG actually was taken up by internal organs, although in low amounts. Following on, Dalmo *et al.* (1994) showed that laminaran and a sulfated analogue of it were taken up by the intestine and were observed to be trapped in organs important in the fish's immune system (kidney, spleen, liver, intestine). A water soluble sulfated laminaran was reported to be present in blood, intestinal tissue and in internal organs in higher amounts than the underivatised laminaran [58]. Experiments on the uptake of VitaStim® (a partly soluble GL) has also been performed [29]. The authors were not able to find any intestinal uptake of neiter the soluble nor particulate part of this substance in channel catfish gut. Based on this finding and the fact that microparticulates of polystyrene were not taken up by the intestinal epithelial cells, it seems that the intestine of salmonids is impermeable to such particles [59].

The tissue- and organ distribution of intravenously injected laminaran in Atlantic salmon and cod (*Gadus morhua*) has also been studied in detail [60, 61]. The main findings were that laminaran was confined to endothelial cells and macrophages situated in kidney, spleen and heart. The salmon kidney and cod heart were shown to be very efficient scavenger organs for this substance.

# 6. Future prospects

As observed in rodents, the therapeutic potential of particulate GLs that are to be injected is hampered by serious side-effects such as granulomatous reaction within the internal organs [10] and the fact that such substances sensitities animals to endotoxin [10, 62]. The fish are extremely insensitive to LPS (also called endotoxin) [63] compared to many mammalian species rendering particulate GL, in this respect, as a good candidate as immunostimulant. Furthermore, particulate GL (MacroGard®) has been shown not to cause such serious intra-abdominal granulomatous lesion, after an i.p injection, as reported to be the case in some mammalian species [64]. As far as side-effects concerned, GLs that form gels when treated with water (lentinan, scleroglucan, shizophyllan etc.) have not been thoroughfully tested. The fully water soluble GLs (laminarin and low molecular weight GLs) have not been observed to cause such lesions. Today, most of the vaccines are based on killed bacteria, bacterin or antigens emulsified in oil [65], and often the vaccine preparation may cause serious side-effects [66]. In order to prevent such non-wanted side-effects, non-oil based adjuvants such as Al-salt, levamisol and GLs could be used. But to date, vaccine preparation containing non-oil adjuvants have not been effective in increasing the survival rate of fish undergoing bacterial infections, such as *A. salmonicida* infection, that result in high mortality rates [66].

If there is a demand for fish that are not "suffering" of intra-abdominal lesions (inflammation) the use of GLs as adjuvants in vaccine preparations may be superior to other. A comparison of the side-effects caused by different GLs should thus be of importance in evaluating the GL of choice. In addition, the efficiency of the different GL adjuvanted vaccine preparations in increasing the survival rate should also be ruled out.

Soluble GLs have apparently a potential to elevate non-specific defence mechanisms in fish larval system, especially to fish that have a long yolk-sac stage (e.g Atlantic halibut). By drinking, osmotically active substances have (low molecular weight) substances may gain access into the yolk-sac larvae [67], and macromolecules such as laminaran have been reported to be absorbed by the fish´s intestine after the yolk-sac larvae were exposed to the substance [68]. Therefore it should be evident that soluble immunostimulants should be considered to increase activity of the non-specific defence mechanisms in fish larvae.

For further litterature regarding the non-specific defence mechanisms in fish, and the use of immunostimulants to modulate them, please find: [69-76].

## ACKNOWLEDGEMENTS

The authur wish to thank dr. Jarl Bøgwald and research fellow Anne L. Myhr for valuable comments and their careful reading of the manuscript. Also the Norwegian Research Council is acknowledged for its financial support.

## References

[1] Bohn J.A. & BeMiller J.N. (1995*Carbohydrate Polymers* 28, 3-14.
[2] Kulicke W.-M., Lettau A. I. & Thielking H. (1997) *Carbohydrate Research* 297, 135-143.
[3] Seljelid R. (1990) In: *Pathogenesis of wound and biomaterial-associated infections*, pp. 107-113. (eds. T. Wadström et al.), Springer Verlag, New York.
[4] Gordon S., Keshav S. & Chung L.P. (1988) *Current Opinion in Immunology* 1, 26-35.
[5] Bayne C. J. (1995). In: *Modulators of immune responses. Phylogeny of regulators of immune responses and immunotoxicants*, pp. 51-53. SOS Publications, Fair Haven, New Jersey, USA, vol.1.
[6] Chihara G., Maeda Y.Y. & Hamuro J. (1982) *International Journal of Tissue and Reaction* IV, 207-225.
[7] Chirigos M.A. (1992) *Thymus* 19 (Suppl. 1), S7-S20.
[8] Kotani S. (1976). *Biken Journal* 19, 9-13.
[9] Kohashi O. (1980) *Infection & Immunology* 29, 70-75.
[10] Bowers G.J., Patchen M.L., MacVittie T.J., Hirsch E.F. & Fink M.P. (1986) *International Journal of Immunopharmacology* 8, 313-321.
[11] Chihara G., Maeda Y.Y., Hamuro J., Sasaki T. & Fukuoka F. (1969) (Berk.) sing. *Nature* 222, 687-688.
[12] Engstad R. E. & Robertsen B. (1995) *Journal of Marine Biotechnology* 3, 203-207.
[13] Seljelid R. & Eskeland T. (1994) *European Journal of Hematology* 51, 267-275.
[14] Sveinbjørnsson B. (1998) Immunomodulation of murine tumors. Studies on cellular mechanisms and mediators. Dr. thesis, Department of Experimental Pathology, Institute of Medical Biology, University of Tromsø, Norway. ISBN 82-7589-086-1.
[15] Czop J.K. & Austen K.F. (1985) *Journal of Immunology* 134, 2588-2593.
[16] Konopski Z., Rasmussen L.-T., Seljelid R. & Eskeland T. (1991) *Scandinavian Journal of Immunology* 33, 297-306.
[17] Ross G.D., Cain J.A., Myones B.L., Newman S.L. & Lachmann P.J. (1987) *Complement* 4, 61-74.
[18] Thornton B.P., Vetvicka V., Pitman M., Goldman R.C. & Ross G.D. (1996). *Journal of Immunology* 156, 1235-1246.
[19] Ross G.D & Vetvicka V. (1993) *Clinical and Experimental Immunology* 92, 181-184.
[20] Ainsworth A.J. (1994. *Veterinary Immunology and Immunopathology* 41, 141-152.
[21] Engstad R.E. & Robertsen B. (1993) *Developmental and Comparative Immunology* 17, 319-330.
[22] Elstad M.R., Parker C.J., Cowley F.S., Wilcox L.A., McIntyre T.M., Prescott S.M. & Zimmerman G.A. (1994) *Journal of Immunology* 152, 220-230.
[23] Muller A., Portera C., Love E., Ensley H., Kelley J., Ha T., Rice P., Orcutt R., Browder W. & Williams D. (1995. *European Journal of Cancer* 31A, 42.
[24] Okazaki M., Adachi Y., Ohno N. & Yadomae T. (1995). *Biological & Pharmaceutical Bulletin* 18, 1320-1327.

[25] Adachi Y., Okazaki M., Ohno N. & Yadomae T. (1997). *Mediators of Inflammation* 6, 251-256.

[26] Adachi Y., Ohno N. & Yadomae T. (1998) *Biological & Pharmaceutical Bulletin* 21, 278-283.

[27] Jørgensen J.B. & Robertsen B. (1995). *Developmental and Comparative Immunology* 19, 43-57.

[28] Tahir A & Secombes C.J. (1996). *Fish & Shellfish Immunology* 6, 135-146.

[29] Ainsworth A. J., Mao C. P. & Boyle C. R. (1994) In: *Models for environmental toxicology, biomarkers, immunostimulators*, (eds. J. S. Stolen and T. C. Fletcher), pp. 67-81. SOS Publications, Fair Haven, NJ, USA, vol.1.

[30] Sveinbjørnsson B. & Seljelid R. (1994) *Veterinary Immunology and Immunopathology* 41, 113-123.

[31] Smedsrød B. & Seljelid R. (1991). *Immunopharmacology* 21, 149-158.

[32] Dalmo R.A. & Seljelid R. (1995) *Journal of Fish Diseases* 18, 175-185.

[33] Jørgensen J.B., Lunde H. & Robertsen B. (1993) *Journal of Fish Diseases* 16, 313-325.

[34] Jørgensen J.B., Sharp G.J.E., Secombes C.J. & Robertsen B. (1993). *Fish & Shellfish Immunology* 3, 267-277.

[35] Brattgjerd S., Evensen Ø. & Lauve A. (1994). *Immunology* 83, 288-292.

[36] Chen D. & Ainsworth A.J. (1992) *Journal of Fish Diseases* 15, 295-304.

[37] Yano T., Mangindaan R.E.P. & Matsuyama H. (1989). *Nippon Suisan Gakkaishi* 55, 1815-1819.

[38] Santarem M., Novoa B., Figueras A. (1997) *Fish & Shellfish Immunology* 7, 429-437.

[38] Yano T., Matsuyama H. & Mangindaan R.E.P. (1991) *Journal of Fish Diseases* 14, 577-582.

[39] Ogier de Baulny, M., Quentel C., Vornier V., Lamour F., Le Gouvello R. (1996). *Diseases of Aquatic Organisms* 26, 139-147.

[40] Verlhac V., Gabaudan J., Obach A., Schüep W. & Hole R. (1996). *Aquaculture* 143, 123-133.

[41] Verlhac V., Obach A., Gabaudan J., Schüep W. & Hole R. (1998) *Fish & Shellfish Immunology* 8, 409-424.

[42] Matsuyama H., Mangindaan R.E.P. & Yano T. (1991) *Aquaculture* 101, 197-203.

[43] Engstad R. E., Robertsen B. & Frivold E. (1992). *Fish & Shellfish Immunology* 2, 287-297.

[44] Efthimiou S. (1996) *Journal of Applied Ichtyology* 12, 1-7.

[45] Siwicki A.K., Anderson D.P. & Rumsey G.L. (1994) *Immunology and Immunopathology* 41, 125-139.

[46] Duncan P.L. & Klesius P.H. (1996). *Journal of Aquatic Animal Health* 8, 241-248.

[47] Dalmo R.A., Bøgwald J., Ingebrigtsen K. & Seljelid R. (1997) *Journal of Fish Diseases* 19, 449-457.

[48] Dalmo R.A., Martinsen B., Horsberg T.E., Ramstad A., Syvertsen C., Seljelid R. & Ingebrigtsen K. (1998) . *Journal of Fish Diseases* (InPress).

[49] Nikl L., Evelyn T.P.T. & Albright L.J. (1993) . *Diseases of Aquatic Organisms* 17, 191-196.

[50] LaPatra S.E., Lauda K.A., Jones G.R., Shewmaker W.S. & Bayne C.J. (1988. *Fish & Shellfish Immunology* 8, 435-446.

[51] Anderson D.P. & Siwicki A.K. (1994) *The Progressive Fish Culturist* 56, 258-261.

[52] Wang W.-S. & Wang D.-H. (1997) *Comparative Immunology, Microbiology and Infectious Diseases* 20, 261-270.

[53] Robertsen B., Rørstad G., Engstad R. & Raa J. (1990*Journal of Fish Diseases* 13, 391-400.

[54] Nikl L., Albright L.J. & Evelyn T.P.T.(1991) *Diseases of Aquatic Organisms* 12,7-12.

[55] Ingebrigtsen K, Horsberg T.E.,Dalmo R. & Seljelid R. (1993) *Aquaculture* 117,29-35.

[56] Samuel M., Lam T.J. & Sin Y.M. (196) *Fish & Shellfish Immunology* 6, 443-454.

[57] Sveinbjørnsson B., Smedsrød B., Berg T. & Seljelid R. (1995) *Fish & Shellfish Immunology* 5, 39-50.

[58] Dalmo R.A., Ingebrigtsen K., Horsberg T.E. & Seljelid R. (1994) . *Journal of Fish Diseases* 17, 579-589.

[59] Dalmo R.A., McQueen Leifson R. & Bøgwald J. (1995) *Journal of Fish Diseases* 18, 87-91.

[60] Dalmo R.A., Ingebrigtsen K., Bøgwald J., Horsberg T.E. & Seljelid R. (1995). *Journal of Fish Diseases* 18, 545-553.

[61] Dalmo R.A., Ingebrigtsen K., Sveinbjørnsson B. & Seljelid R. (1996) *Journal of Fish Diseases* 19, 129-136.

[62] Lazar G. & Agarwal M.K. (1982) *Biochemistry and Medicine* 28, 310-318.

[63] Berczi I., Bertok L. & Bereznai T. (1966) . *Canadian Journal of Microbiology* 12, 1070-1071.

[63] Midtlyng P.J., Reitan L.J., Lillehaug A. & Ramstad A. (1996) *Fish & Shellfish Immunology* 6, 599-613.

[64] Anderson D.P. (1992) *Annual Review of Fish Diseases* 2, 281-307.

[65] Midtlyng P.J. (1998) Evaluation of furunculosis vaccines in Atlantic salmon. Experimental and field studies for assessment of protection and side-effects. Dr.thesis, Norwegian College of Veterinary Medicine, Oslo, Norway. ISBN 82-90550-29-4.

[67] Tytler P. & Blaxter J.H.S (1988] *Journal of Fish Biology* 32, 493-494.

[68] Strand H.K. & Dalmo R.A. (1997) *Journal of Fish Diseases,* 20, 41-49.

[69] Ellis A.E. (1981In: *International Symposium on Fish Biologics: serodiagnostics and vaccines. Developments in biological standardization*, proceedings of a symposium organized by the International Association of Biological Standardization, vol. 49 pp 337-352. (Ed. W. Hennessen), S. Karger, Basel, Switzerland.

[70] Alexander J.B. & Ingram G.A. (1992] *Annual Review of Fish Diseases* 2, 249-279.

[71] Secombes C.J. (1994). *Fish & Shellfish Immunology* 4, 421-436.

[72] Secombes C.J. (1996). In: *The fish immune system: organism, pathogen and environment,* pp 63-103. (Eds. G. Iwama and T. Nakanishi), Academic Press Inc. San Diego, California, USA,

[73] Raa J. (1996) *Reviews in Fisheries Science* 4, 229-288.

[74] Yano T. (1996). In: *The fish immune system: organism, pathogen and environment.* pp 105-157. (Eds. G. Iwama and T. Nakanishi), Academic Press Inc. San Diego, California, USA.

[75] Anderson D.P (1997). In: *Fish Vaccinology, International Association of Biological Standardization*: International symposium in vaccinology, VESO, Oslo, Norway, 5.-7. June 1996. pp 257-265. (Eds. R. Gudding *et al.*), Karger, Basel, Switzerland.

[76] .Dalmo R.A., Ingebrigtsen K. & Bøgwald J. (1997) *Journal of Fish Diseases* 20, 241-273.

Chapter 9

# Extractability of cell wall-related polysaccharides with potential bioactivity and their persistence during fermentation

J.A.ROBERTSON, P.RYDEN and S.G.RING

*Department of Biochemistry, Institute of Food Research, Norwich Research Park, Colney, NORWICH NR4 7UA, UK*

Key words: cell walls, fermentation, non-starch polysaccharides,

Abstract: Non-starch polysaccharides (NSP) make an important contribution to the diet but mechanisms for their health promoting properties as immunostimulants has received limited attention. NSP, from fruit and vegetable sources in particular, are rich in pectic polysaccharides and have the potential to act as immunostimulants during gut transit. To clarify the potential for bioactivity through diet, the contribution of NSP has been considered in terms of the exposure levels to pectic polysaccharides in foods (~4g/day) and their expected behaviour and persistence during digestion and fermentation. Structural features of polysaccharides associated with bioactivity can be identified in fruits and vegetables but at low exposure levels. Chronic exposure through diet and persistence of polysaccharides during fermentation may be important to trigger immunostimulation and need to be considered when interpreting the potential for polysaccharide-based bioactivity through diet.

## 1. Introduction

Polysaccharides are universal constituents of living organisms and a major component of many foods. Their contribution to food is mainly as starch but the non-starch polysaccharides (NSP) as mainly plant cell wall polysaccharides or dietary fibre, have an identified role in health promotion. In terms of diet bioactivity can be considered as the positive health response



*B.S. Paulsen (ed.), Bioactive Carbohydrate Polymers,* 107–119.

generated and maintained at acceptable dosages of polysaccharides (NSP) in foods. Mechanisms for health promoting properties remain unclear [1] but bulking effects, association with other dietary components, eg phytochemicals and fermentation to short chain fatty acids have each been suggested to play a role [2]. The immunostimulating properties of diet-derived polysaccharides has received limited attention, despite their chronic contribution to the diet and advice to increase consumption for health benefit.

In general, bioactivity involving polysaccharides has been associated with a relatively high molecular weight polysaccharide fraction, with fungal polysaccharides associated more with anticarcinogenic effects and uronic acid rich polysaccharides associated with immunostimulation [3]. The dosage given has often also used 'isolated' polyaccharides, at relatively high concentrations, to obtain a localised health response, eg wound healing [4] and it is unknown how this relates to threshold concentrations which may be required for immunostimulation. Diet-derived polysaccharides, from fruit and vegetable sources in particular, are rich in pectic polysaccharides and hence uronic acids which may act as immunostimulants during gut transit. To clarify the potential for bioactivity from diet-derived polysaccharides requires consideration of both the exposure levels to the bioacive agent and the structural similarities between diet-derived and isolated bioactive polysaccharides. Chronic exposure, as the daily intake or dosage of polysaccharides in foods, will be important, as will the behaviour of the constituent polysaccharides during digestion and gut transit, since this may affect the release or persistence of structures important for functional behaviour as an immunostimulant.

## 2. Presence of NSP in foods

### *Sources and Amount*

The major contributor of NSP to the diet is the plant cell wall from vegetables, fruits and cereals, although there is increasing use of 'fibre concentrates', eg resistant starches and fructo-oligosacharides in food processing. A general classification of sources of NSP and range of amounts expected to be provided/100g fresh weight is illustrated in Figure 1. Fruits and vegetables can be considered to provide between 1.5 - 2.5g/100g fresh weight [5]. Cereals contribute more NSP to the diet but amount does depend largely on the extent of grain refining, eg 72% extraction flour contains 3.2% NSP but wholemeal flour has 9.8% NSP [5]. Legume seeds have an apparently high NSP content but after rehydration and cooking, NSP content/100g fresh weight is similar to vegetables and cereals. Fibre concentrates, eg apple pomace, are by definition NSP-enriched but incorporation into foods reduces their consumption level similar to that of fruits, vegetables and cereals. Fibre isolates are generally extracted polysaccharides, eg alginates and 'pectins', used as emulsifying or texturising agents during food processing. Their contribution to the diet is small but potentially significant if they have a demonstrated bioactivity.

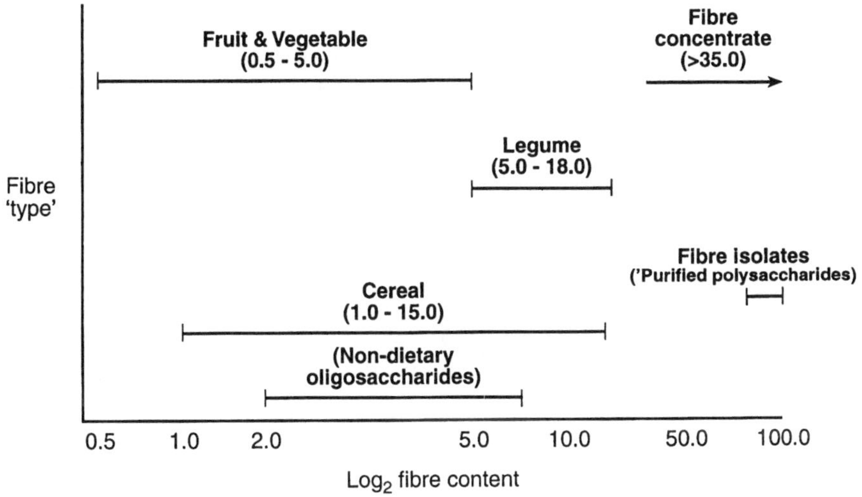


Fig. 1. Dietary fibre content, as nonstarch polysaccharides, expected for major plant-based food sources or 'types' (g fibre/100g food product).

Daily intake of NSP is derived from a combination of the above sources and for a typical 'European diet' is between 15 and 20g/day. A North European diet tends to contribute more from cereal sources, 65% cereal and 35% fruit and vegetable, whilst a Mediterranean diet has a greater contribution from fruits and vegetables. On the basis of a North European diet then around 7g NSP/day is derived from fruits and vegetables and around 13g from cereals. This provides guidelines to assess exposure limits to diet-derived polysaccharides and to interpret bioactivity in terms of dietary response.

***Composition***

As well as variation in amount of NSP/100g fresh weight there are differences in the relative contribution of component polysaccharides (Table 1). Thus pectic polysaccharides, as the sum of galacturonic acid, arabinose, galactose and rhamnose, are the major component of NSP from commonly available fruits and vegetables and contibute around 60% to the NSP from the plant cell wall [6]. Galacturonic acid can be considered the major sugar present in pectic polysaccharides and may contribute around 50% of the total pectic polysaccharide fraction. Pectic polysaccharides are usually present in only trace amounts in cereal grain fractions. Cellulose contributes around 30% of NSP from fruit and vegetable sources but less from cereal grain fractions. The hemicelluloses are the major polysaccharides present in

| Source | NSP/100g Fresh weight | % Cellulose | % Pectic Polysaccharide | % Hemicellulose |
|---|---|---|---|---|
| Carrot | 2.4 | 33 | 63 | 4 |
| Onion | 2.0 | 34 | 62 | 4 |
| Cauliflower | 1.9 | 30 | 57 | 13 |
| Potato | 1.2 | 31 | 58 | 11 |
| Mung bean | 12.2 | 38 | 44 | 18 |
| Wheat (bran) | 35.0 | 18 | Trace | 82 |
| Oats (bran) | 12.8 | 10 | Trace | 90 |

Table 1 Nonstarch polysaccharide (NSP) content for a range of commonly available vegetable sources and cereal brans [5] and the relative distribution of polysaccharide 'types' within each source.

cereals, mainly as arabinoxylans but mixed linkage β-glucans are an important component in some cereals, eg oats, in relation to dietary effects or bioactivity. On the basis of a 7g/day intake of NSP from fruits and vegetables then dosage of pectic polysaccharides, supplied at mealtimes, will be around 4g/day and exposure of structural features which may be important for bioactivity will be related to the behaviour of meals during digestion and gut transit.

## 3. Intake of NSP and dietary reponse

Dietary response, eg effects on serum cholesterol, varies with the fibre source and there can be apparent discrepencies in response. Thus in one study [7] carrot, supplemented as 4g fibre/person/day as raw carrot, was shown to be effective in decreasing serum cholesterol whilst in a corresponding study carrot, supplemented at above 15g fibre/person/day as raw or processed carrot, showed no effect [8]. This would appear to argue against fibre having a bioactive role in the diet but in the second study [8] subjects were active athletes and hence probably limited in their ability to respond to the test diet compared to subjects in the first study [7]; ie bioactivity through diet should be considered in conjunction with general lifestyle and the potential to improve general health through diet. Vegetable sources related to onions tend to decrease serum cholesterol and, although not tested for cauliflower, similar effects are found for the cabbage family. Legumes, like mung bean, are also considered to be effective at reducing serum cholesterol but potatoes, like wheat bran, have no demonstrated effect.

| Source | NSP 'type' | Polysaccharide | Major 'types' | Dietary effect |
|--------|-----------|----------------|---------------|----------------|
| Carrot | soluble | Pectic | Arabinogalactan | Decrease? |
| Onion | soluble | Pectic | Galactan/ Arabinogalactan | Decrease |
| Cauliflower | soluble | Pectic | Arabinan/ Arabinogalactan | Decrease? |
| Potato | soluble | Pectic | Galactan/ Arabinogalactan | None |
| Mung bean | soluble | Pectic | Arabinan/ Arabinogalactan | Decrease |
| Wheat (bran) | insoluble | Arabino-xylan | Ara:Xyl ~ 1:1 Ara:Xyl ~ 1:2, Acidic | None |
| Oats (bran) | soluble | β glucan | Mixed linkage | Decrease |

Table 2   Classification systems for nonstarch polysaccharides from a range of commonly available vegetable sources and cereal brans and in relation to bioactivity or observed dietary effect on serum cholesterol.

Attempts have been made to relate dietary effect to 'type' of NSP as soluble fibre but 'soluble fibre' is very dependent on the method of analysis [9] and does not correlate with dietary response (Table 2). Similarly dietary response to pectic polysaccharides is not consistent, even when classified on the basis of their contribution as galactan, arabinan and arabinogalactan type pectic polysaccharides and mixed linkage β-glucan can give a response similar to pectic polysaccharides. Thus, there is an apparent discrepancy in the ability to relate bioactivity to polysaccharide type but as yet no account has been taken of the presence of structural features in polysaccharides associated with bioactivity.

## 4. Structural features and bioactivity

Polysaccharides with ascribed bioactivity are present in a wide range of plant food sources and include rhamnogalacuronan II (RG-II) [10] , arabinogalactans [11,12] and βglucans[13]. The arabinogalactans with ascribed bioactivity are Type II arabinogalactans [14], containing (1->3,6) linked galactose, and both arabinogalactan side chains from pectic polysaccharides [11] and arabinogalactan proteins [12] demonstrate bioactivity. Activity of β-glucan is associated with the (1->3) linkage [13]. For both arabinogalactans and β-glucan, associated sugars or substitutions through C6 are thought to be important to enhance solubility, although the helical structure dictated by the (1->3) linkage [15] will expose C6 substitutions at the surface of the helix. This may identify an important link between polysaccharide structure and its functional behaviour but our understanding of the mechanisms involved in expressing bioactivity remain limited. Similarly the dimerisation of RG-II appears to be important for bioactivity [10] but again the mechanism of expression remains unknown.

## 5. 'Bioactive' structural features and NSP

Rhamnogalacturonan II and type II arabinogalactans can be identified in a wide range of cell wall preparations but, as can be shown for the arabinogalactans, their presence can depend on the method of cell wall preparation. Thus, in a carrot alcohol-insoluble residue (AIR) [16] and in a water-insoluble residue from carrrot pulp [17] (1->3,6) linked galactose was present but a hot water extract of carrot AIR removed a highly esterified acidic arabinogalactan [16]. A major contributor to the water extract was Type 1 arabinogalactan [14], ie containing (1->4)-linked galactose, but some (1->3,6)-linked galactose was also extracted. The 1M KOH extract also contained (1->3,6)-linked galactose, suggesting there may be 2 structurally distinct arabinogalactans associated with carrot AIR, arabinogalactan-protein and arabinogalactan associated pectic polysacharides. Indeed pectic polysaccharides extracted from carrot tissue culture have associated regions of arabinogalactan [18]. However, the linkage was not always detected in AIR after pronase treatment [19]. Similarly, (1->3,6) linked galactose can be identified in the water-insoluble residue from onion [17] but is absent from cell walls prepared using detergents and phenol / acetic acid / water (PAW) extraction [20]. Also, whilst present in the water-insoluble residue from cabbage [17] and cabbage AIR [21], in potato the linkage is absent in both water-insoluble cell wall residues [17] and in PAW extracted cell wall material [22]. However, in apple, (1->3,6) linked galactose can be identified in both AIR residues [23] and in extracted and purified cell wall pectic polysaccharide material [24,25]. These observations suggest that the (1->3,6) linked galactose may be present as an arabinogalacan protein, eg water soluble, pronase degradable and PAW extractable, but also that the linkage can be present bound within the cell wall matrix, probably as a component of the modified hairy regions of pectic polysaccharides. Loss of soluble arabinogalactan, and possibly RG-II, during cell wall preparation, would be consistent with the extraction of 'bioactive agents' during the preparation of Kampo medicines and demonstrates how potentially bioactive agents may be lost from NSP by extraction or processing conditions.

## 6. Extractability of pectic polysaccharides

Sequential extraction of cell wall or NSP preparations in a series of aqueous solvents [26] can be used to isolate and monitor associations between polysaccharides within the cell wall matrix (Table 3). Thus, pectic polysaccharides are expected to be extracted under relatively mild conditions, whilst a more severe alkali treatment is expected to be required to disrupt associations involving the hemicelluloses and hence to remove hemicelluloses and leave a residue of $\alpha$–cellulose. However, the observed extractability of pectic polysaccharides can be very different from expected when using purified cell wall material prepared from raw vegetables (Table 3). Whilst a third of pectic polysaccharides may be extracted from carrot using CDTA and a large proportion is extractable in carbonate, with little extractable along with the hemicelluloses, around a third remains associated with the $\alpha$–cellulose residue. Similarly, the profiles of pectic polysaccharide extraction from onion, cauliflower, potato and mung bean differ. For example, almost 50% of the pectic polysaccharides in potato remain associated with the $\alpha$–cellulose residue whilst little pectic polysaccharide is released from mung bean cotyledon prior to extraction with 4M KOH and again a large proportion remains associated with the $\alpha$–cellulose residue. This has implications for the release from the cell wall matrix of potentially bioactive agents during digestion. Thus, the more water soluble polysaccharide material, eg arabinogalactan-based and possibly RG-II, may be released during digestion in the stomach and small intestine and stomach acidity may promote release of some CDTA soluble pectic polysaccharide. However, a large scale solubilisation of pectic polysaccharide would appear unlikely in the stomach and small intestine, although amounts can be boosted by cooking and processing.

When pectic polysaccharides are heat treated they can undergo a partial degradation [29], which increases their extractability. Thus, after extrusion processing pectic polysaccharides from mung bean become more readily extractable in water, CDTA and carbonate [30], and similar effects can be obtained using highly insoluble pea hulls [31] and even extractability of wheat bran arabinoxylans can be enhanced following extrusion treatment [32]. The effect can be illustrated using carrot AIR (Table 4). In raw carrot around 10% of the polysaccharide in the AIR is water soluble and, as noted for the purified carrot cell wall (Table 3) a large proportion is intimately associated with the $\alpha$–cellulose residue. After cooking the water-extractability increases, as does the extractability into CDTA and carbonate,

| Cell wall source | | Carrot | Onion | Cauliflower | Potato | *Mung bean |
|---|---|---|---|---|---|---|
| *% Total polysaccharide* | | *63* | *62* | *57* | *58* | *44* |
| **Extract** | **(Expected)** | **% Extracted** | | | | |
| 50mM CDTA | Pectic polysaccharide | 33 | 16 | 19 | 22 | 5 |
| 50mM $Na_2CO_3$ | | 29 | 36 | 49 | 25 | 7 |
| 0.5M KOH | Hemicelluloses | 2 | 15 | 12 | 1 | 1 |
| 1.0M KOH | | 1 | 4 | 2 | 3 | 2 |
| 4.0M KOH | | Trace | 4 | 2 | 2 | 52 |
| Residue | Cellulose | 35 | 24 | 15 | 46 | 33 |

Table 3 Extractability of pectic polysaccharides from a range of commonly available vegetable sources. (For extractability procedures see [25]. CDTA = trans-1,2-cyclo-hexanediamine-N,N,N',N'-tetraacetate. Carrot - unpublished data; Onion - [20]; Cauliflower - [27]; Potato - [22]; Mung bean (cotyledon) - [28]).

| Sample | Raw | Cooked |
|---|---|---|
| *% Total polysaccharide* | *66* | *62* |
| **Extract** | **% Extracted** | |
| Water | 13 | 24 |
| 50mM CDTA | 19 | 36 |
| 50mM $Na_2CO_3$ | 40 | 20 |
| 0.5M KOH | 5 | 3 |
| Residue | 22 | 11 |

Table 4. Effects of cooking on the extractability of pectic polysaccharides from AIR prepared from carrot. (Raw carrot and raw carrot boiled 20 min in distilled water were each used to provide AIR as 'raw' and 'cooked' preparations. Each preparation was extracted in aqueous solvents [25] as shown.)

at the expense of pectic polysaccharide associated with the $\alpha$-cellulose residue. This increased lability of pectic polysaccharides could be important in extracting bioactive polymers, like the use of hot water extracts used in the preparation of Kampo medicines, and important in explaining inconsistencies in response to NSP in the diet. Changes in the profile of pectic polysaccharide structures or linkage patterns following cooking and which may be consistent with proposed bioactivity have yet to be determined. However, it is clear that extractability differs between NSP sources and can be modified through cooking. This will affect exposure of polysaccharides to the gut mucosa and also affect the fermentability, ie extensive breakdown, of the cell wall matrix.

## 7. Fermentation

Fermentation of polysaccharides occurs in the large intestine (colon) and involves the release of polysaccharides from the cell wall matrix or food residues and their subsequent acidogenic breakdown to produce short chain fatty acids, eg acetate, propionate and butyrate. Rate of fermentation of fibre sources in the large intestine, is relatively rapid [33]. The extent of fermentation, or persistence of insoluble substrate, varies with fibre source but, in the caecum or proximal colon, is generally reached within 18 hours [33], ie fermentability is complete within the transit time expected for colonic contents. In terms of bioactivity it is perhaps more relevant to consider the presence of soluble polysaccharides and their persistence in the large intestine in terms of dose effective for bioactivity.

As shown in Figure 2 for an *in* vitro fermentation system the acidogenic potential for glucose is achieved within 2 hours whereas with extracted polysaccharide isolates, the acidogenic potential persists for up to 12 hours and 'cooking' can affect the rate of fermentation. With glucose only acidogenesis is being measured whereas with the polysaccharides their degradation to oligomers and monosaccharides and subsequent acidogenesis is being measured. Hence polysaccharide-related material would be expected to persist for several hours in the colon. In terms of releasing bioactive polysaccharides from NSP in the colon then concentration of bioactive polysaccharide has to be considered in terms of their persistence during fermentation, turnover and accumulation as solubilised polysaccharide is released and time of exposure required to elicit a bioactive response. Certainly polysaccharides can persist for several hours in the colon and concentrations can be envisiged within the range used to demonstrate immunostimulating properties (10μg/ml) [10]. However, persistence in the colon can depend on polysaccharide source and, although chronic exposure to polysaccharides can be claimed, the exposure in the colon will be intermittent and related to dietary habits (food source, amount eaten, meal frequency) and gastric emptying and transit time in the small intestine.

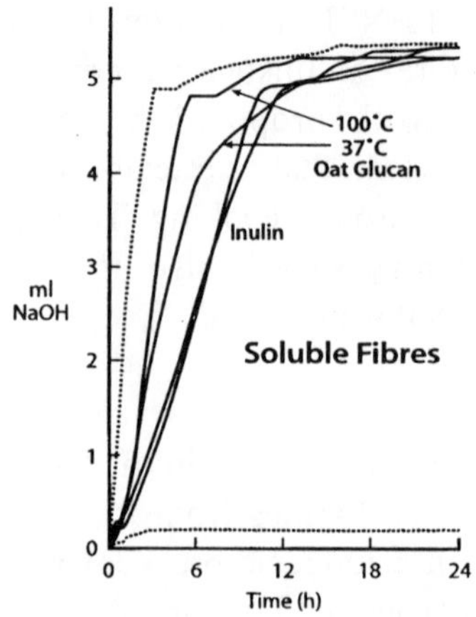


Fig.2 Relative fermentability of soluble polysaccharide isolates measured as net acid production. (1.7g substrate/0.7% concentration was fermented at 37$^0$C under anaerobic conditions using a 10% faecal inoculum and acidity was automatically titrated using 2M NaOH to maintain pH = 6.5).

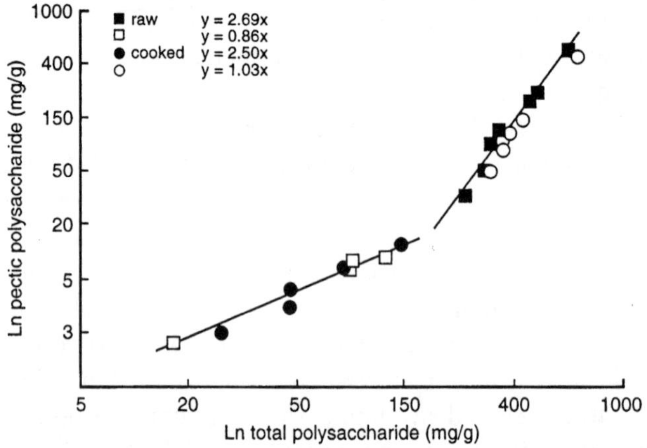


Fig. 3 Fermentability of pectic polysaccharides from raw and cooked carrot relative to the fermentability of the total polysaccharides or NSP in each sample. (For no effect of polysaccharide source then relative rate ~1; for preferential loss of polysaccharide source then relative rate >1).

Pectic polysaccharides tend to be preferentially lost during fermentation and this has implications for the release and exposure to pectic-related bioactive polysaccharides. For example, with carrot (Figure 3), if pectic polysaccharides are fermented at a rate similar to other polysaccharides then the expected relative rate of fermentation equals 1, but the observed rate for the bulk of the pectic polysaccharides is ca. 2.5 times and there was no difference between raw and cooked carrot. A minor component of the pectic polysaccharides is fermented at a rate similar to polysaccharides overall. A more rapid rate of solubilisation, without affecting the rate of acidogenesis, would result in a higher concentration of potentially bioactive polysaccharides in the proximal colon but conversely probably a lower concentration towards the distal colon. How rates of fermentation, release and persistence of potentially bioactive polysaccharides in the colon relate to immunostimulation of the colonic mucosa has yet to be determined.

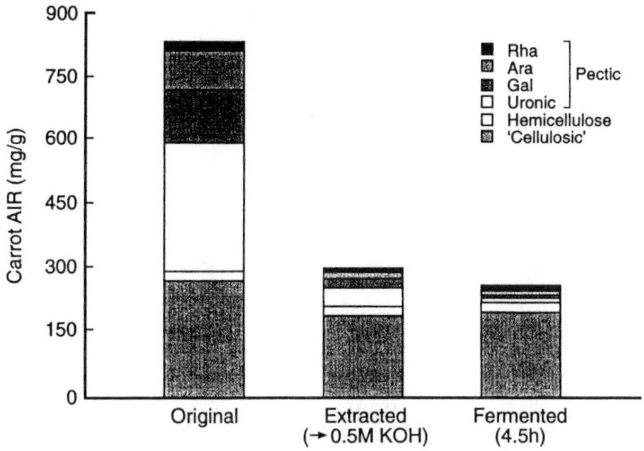


Fig. 4 Composition of carrot AIR before and after partial fractionation using aqueous solvents [25] and after a controlled fermentation to release pectic polysaccharides.

Our current approach to looking for bioactivity from diet-derived polysaccharies is to exploit the behaviour of NSP during sequential extraction using aqueous solvents and to compare this extractability to losses during fermentation. As shown in Figure 4, sequential extraction up to and including 0.5M KOH removes the bulk of the pectic polysaccharides and leaves a residue similar in composition to a sample fermented *in vitro* for around 4.5 hours (~ 70% polysaccharide loss). The underlying assumption is that the extract can be used to 'mimic' solubilisation through fermentation and to obtain extracts for testing bioactivity before and as a result of fermentation. However, as with other workers involved in identifying bioactive polysaccharides related to health promotion, bioassays are required to test and monitor bioactivity at expected or proposed dietary exposure levels. If exposure levels and response can be rationalised to dietary or pharmaceutical dosage then this will help resolve the role of polysaccharides as immunostimulants and hence the mechanism(s) involved in eliciting a positive response to NSP in the diet.

## References


[1] Eastwood, M.A. (1992) *Ann. Rev. Nutr. 12*, 19-35.

[2] Cummings, J.H., Bingham, S.A., Heaton, K.W. & Eastwood, M.A. (1992) *Gastroenterol. 103*, 1783-1789.

[3] Srivastava, R., & Kulshreshtha, D.K. (1989) *Phytochem., 28,* 2877-2883.

[4] Williams, D.L., Mueller, A. & Browder, W. (1995). *J. Endotoxin Res., 2,* 203-208.

[5] Holland, B., Welch, A.A., Unwin, I.D., Buss, D.H., Paul, A.A. & Southgate, D.A.T. (1991) McCance & Widdowson's *The Composition of Foods*, 5th ed. Royal Soc. Chem. and MAFF, London.

[6] Selvendran,R.R., Stevens, B.J.H. & DuPont, M.S. (1987) *Adv. Food Res. 31*, 117-209.

[7] Robertson, J.A., Brydon, W.G., Tadesse, K., Wenham, P., Walls, W. & Eastwood, M.A. (1979) *Am. J. Clin. Nutr. 32*, 1889-1992.

[8] Wisker, E., Schweizer, T.F., Daniel, M. & Feldheim, W. (1994) *Br.JNutr.72*,579-599.

[9] Marlett, J.A., Chesters, J.G.,, Longacre, M.J. & Bogdanske, J.J. (1989) *Am. J. Clin. Nutr. 50*, 479-485.

[10] Shin, K.S., Kiyohara, H., Matsumoto, T. & Yamada, H. (1998) *Carbohydr. Res. 307*, 97-106.

[11] Yamada, H. (1994) *Carbohydr. Polymers 25*, 269-276.

[12] Clarke, A.E., Anderson, R.L. & Stone, B.A. (1979) *Phytochem. 18*, 521-540.

[13] Browder, W., Sherwood, E., Williams, D., Jones, E., McNamee, R. & DiLuzio, N. (1987) *Am. J. Surgery 153*, 25-32.

[14] Aspinall, G.O. (1973) In: *Biogenesis of Plant Cell Wall Polysaccharides.* (ed. F. Loewus), Academic Press, NY p.95.

[15] Fincher, G.B., Stone, B.A. & Clarke, A.E.(1983) *Ann. Rev. Plant Physiol. 34,* 47-70.

[16] Stevens, B.J.H. & Selvendran, R.R. (1984) *Carbohydr. Res. 128*, 321-333.

[17] Schols, H.A. & Voragen,A.J.G. (1994). *Carbohydr. Res. 256*, 83-95.

[18] Kikuchi, A., Edashige, Y., Ishii, T., Fujii, T. & Satoh, S.(1996) *Planta 198*, 634-639.

[19] Massiot, P., Rouau, X. & Thibault, J-F. (1989) *Carbohydr. Res. 172*, 229-242.

[20] Redgwell, R.J. & Selvendran, R.R. (1986) *Carbohydr. Res. 157*, 183-199.

[21] Stevens, B.J.H. & Selvendran, R.R (1980) *J.Sci. Food Agric. 31*, 1257-1267.

[22] Ryden. P. & Selvendran, R.R. (1990) *Carbohydr. Res. 195*, 257-272.

[23] Stevens, B.J.H. & Selvendran, R.R. (1984) *Carbohydr. Res. 135*, 155-166.

[24] Schols H.A., Posthumus, M.A. & Voragen,A.J.G.(1990)*Carbohydr.Res.206*,117-129.

[25] Renard, C.M.G.C., Voragen, A.J.G., Thibault, J-F. & Pilnik, W. (1991) *Carbohydr. Polymers 16*, 137-154.

[26] Selvendran, R.R. & Ryden, P.(1990). In: *Methods of Plant Biochemistry 2*, (Carbohydrates), Academic Press, NY. pp.549-579.

[27] Femenia, A. (1995) Study on the potential of vegetable processing by-products as sources of dietary fibre: effects of the tissue maturation and processing on component

cell wall polysaccharides. Ph.D Thesis: L'Universite de Bretagne Occidentale, Brittany, France. (in english)

[28] Gooneratne, J., Needs, P.W., Ryden, P. & Selvendran, R.R. (1994). *Carbohydr. Res. 265*, 61-77.

[29] Barret, A.J. & Northcote, D.H. (1965). *Biochem. J. 97*, 617-627.

[30] Gooneratne, J. Majsak-Newman, G., Robertson, J.A. & Selvendran, R.R. (1994) *J. Agric. Food Chem. 42*, 605-611.

[31] Ralet, M.C., Saulnier, L. & Thibault, J-F. (1993) *Carbohydr. Polymers 20*, 25-34.

[32] Ralet, M.C., Thibault, J-F. & DellaValle, G. (1991)*J. Cereal Sci. 11*, 249-259.

[33] Robertson, J.A., Murison, S.D. & Chesson, A. (1987) *J. Nutr. 117*, 1402-1409.

Chapter 10

# Pectic hairy regions of lemon fruits: a polysaccharide with potential bioactivity?

J.M. ROS[a], H.A. SCHOLS[b], J. LAENCINA[a] and A.G.J. VORAGEN[b]
*[a]CEBAS-CSIC & University of Murcia, Department of Food Technology, 30100 Murcia, Spain*
*[b]Wageningen Agricultural University, Department of Food Technology & Nutritional Sciences, Food Science Group, 6703 HD Wageningen, The Netherlands*

Key words: Hairy regions, pectin, lemon, bioactivity

Abstract: The hairy regions of lemon pectins from both albedo and juice were isolated and further characterised. Pectic hairy regions and the so called modified hairy regions represent high molecular weight fragments of rhamnogalacturonans with varying amounts of arabinose- and/or galactose-rich side chains. Similarities have been found for lemon pectin side chains when compared to the type of specific pectic structures stated as necessary for bioactivity. Following this approach, the potential bioactivity of pectic hairy regions of lemon fruits is discussed.

## 1. Introduction

Bioactivities observed for pectins and related polysaccharides have been reported recently and may depend on their chemical fine [1,2]. Many of the bioactivities mentioned for pectins may have a relationship with complex structures within the ramified rhamnogalacturonan region of pectins from various sources. Even if native pectin has no activity, chemical and enzymatic modified pectin may provide useful products having pharmacological properties [1].

Lemon peels are of increasing importance as a source for the industrial manufacture of pectins [3]. Like pectin from other sources, lemon pectin has been reported to be constituted of homogalacturonans (smooth regions) and more complex structure of rhamnogalacturonans rich in arabinan, galactan



*B.S. Paulsen (ed.), Bioactive Carbohydrate Polymers,* 121–128.

and arabinogalactan side chains (hairy or ramified regions) [4,5,6,7] Homogalacturonan segments from pectins are active as sequestering agent for heavy metals and binding of cholesterol, lipids and bile acids [8,9,10,11]. In this contribution we discuss on potential bioactivity of the hairy regions of pectins from lemon peel and juice.

## 2. Experimental

*Preparation of pectin fractions from lemon albedo.*—Lemon albedo was sequentially extracted to yield a pectin soluble in chelating agents (ChSS), a pectin soluble in diluted alkali (ASS), and a Residue [6].

*Preparation of a pectin fraction from lemon juice.*—Lemon juice was dialysed and lyophilised to yield a pectic polymeric fraction.

*Isolation of pectic hairy regions.*—Pectin fractions from lemon albedo and juice were incubated with a combination of polygalacturonase and pectinesterase according to Schols *et al.* [12]. The isolation of the hairy region populations was performed by size-exclusion chromatography of the digests.

*Isolation of modified hairy regions.*—Lemon albedo and juice were incubated with Rapidase C600. The isolation of the modified hairy region fractions was performed by cross-flow filtration [13].

*Analytical methods.*—Uronic acids, neutral sugars, degrees of methyl esterification and acetylation, and linkage composition were determined according to Ahmed and Labavitch [14], Blakeney *et al.* [15], Voragen *et al.* [16] and Schols *et al.* [17], respectively.

## 3. Results

*Characterisation of pectic hairy regions.*—In most of the fractions neutral sugar residues were present in higher proportions than the uronic acid residues (Table 1, 2 and 3). The most important neutral sugar residue was arabinose, in addition to galactose and minor amounts of rhamnose. The rhamnose to galacturonic acid ratio varied between 0.1 and 0.3. Some fractions were shown to have higher contents of xylose and glucose.

The methyl esterification of ChSS HR-I (highest Mw fraction of HR as isolated by SEC) is much lower (OMe = 25) than for the lower Mw HR-fraction (ChSS HR-II; OMe = 55). The same is true for the acetyl contents in these I and II hairy region fractions: 22% and 41% respectively.

**Table 1** *Sugar composition (mol%) of the hairy regions from lemon albedo pectins obtained after degradation by PG and PE*

| sugar | ChSS Pop. I | ChSS Pop. II | ASS Pop. I | ASS Pop. II | Residue |
|---|---|---|---|---|---|
| Rhamnose | 4 | 5 | 3 | 5 | 2 |
| Arabinose | 59 | 23 | 58 | 28 | 53 |
| Xylose | 1 | 7 | 5 | 28 | 2 |
| Mannose | 1 | 13 | 1 | 1 | tr |
| Galactose | 21 | 11 | 19 | 8 | 33 |
| Glucose | 1 | 9 | tr | 1 | 1 |
| GalAcid | 15 | 31 | 16 | 29 | 10 |
| OMe | 25 | 55 | | | |
| OAc | 22 | 41 | | | |

**Table 2** *Sugar composition (mol%) of the hairy regions from lemon juice pectin obtained after degradation by PG and PE*

| sugar | HR | Pop. I | Pop. II | Pop. III |
|---|---|---|---|---|
| Rhamnose | 2 | 4 | 4 | tr |
| Arabinose | 10 | 16 | 20 | 2 |
| Xylose | 1 | 6 | 3 | 1 |
| Mannose | 1 | 9 | 6 | 6 |
| Galactose | 14 | 22 | 30 | 2 |
| Glucose | 3 | 13 | 6 | 2 |
| GalAcid | 68 | 29 | 30 | 85 |

**Table 3** *Sugar composition (mol%) of the modified hairy regions from lemon albedo and juice obtained after enzymatic liquefaction*

| sugar | Lemon albedo | Lemon juice |
|---|---|---|
| Rhamnose | 5 | 5 |
| Arabinose | 56 | 10 |
| Xylose | 1 | 1 |
| Mannose | 1 | 6 |
| Galactose | 8 | 25 |
| Glucose | 1 | 6 |
| GalAcid | 28 | 46 |
| OMe | 17 | 39 |
| OAc | 25 | 13 |

**Table 4** *Glycosidic linkage composition of the neutral sugar fraction of the modified hairy region isolated from lemon albedo*

| sugar | Glycosidic linkage composition[a] | |
|---|---|---|
| Rhamnose | T-Rha*p*[b] | 1.1 (15) |
| | 1,2-Rha*p* | 1.6 (22) |
| | 1,2,4-Rha*p* | 3.6 (50) |
| | 1,2,3,4-Rha*p* | 0.9 (13) |
| | Total | 7.2 |
| Arabinose | T-Ara*f* | 16.0 (20) |
| | 1,5-Ara*f* | 53.9 (66) |
| | 1,3,5-Ara*f* | 4.0 (5) |
| | 1,2,5-Ara*f* | 2.9 (4) |
| | 1,2,3,5-Ara*f* | 4.5 (6) |
| | Total | 81.3 |
| Xylose | T-Xyl*p* | 1.4 (67) |
| | 1,3-Xyl*p* | 0.7 (33) |
| | Total | 2.1 |
| Galactose | T-Gal*p* | 3.2 (46) |
| | 1,6-Gal*p* | 0.7 (10) |
| | 1,4-Gal*p* | 1.8 (26) |
| | 1,3-Gal*p* | 0.4 (6) |
| | 1,2,4-Gal*p* | nd |
| | 1,3,4-Gal*p* | nd |
| | 1,3,6-Gal*p* | 0.9 (13) |
| | Total | 7.0 |
| Glucose | T-Glc*p* | 0.4 (24) |
| | 1,4-Glc*p* | 1.3 (77) |
| | Total | 1.7 |
| Ratio terminal/branching | | 1.01 |

[a]Linkage types in mol %; numbers in brackets indicate percentage of each linkage in a sugar residue.
[b]1,2-linked Rha, etc. T connotes a terminal residue.

The hairy regions from lemon juice pectin had a higher galacturonic acid content than those from lemon albedo pectin. As a consequence the rhamnose to galacturonic acid ratio and the amount of arabinose residues were lower in hairy regions from lemon juice pectin. The same is true for the modified hairy region fraction from lemon juice isolated using a technical enzyme preparation when compared to the modified hairy regions from lemon albedo pectin.

*Linkage composition.*—The linkage composition was determined by methylation analysis for the modified hairy region from lemon albedo, since only this fraction was obtained in sufficient quantities. The sugar composition as determined before (Table 3) and after (Table 4) methylation analysis was similar. Table 4 only includes the glycosidic linkage composition of the neutral sugar fraction of the modified hairy region from lemon albedo. When the sugar composition of the permethylated samples are corrected for the galacturonic acid content (28 mol%), the data is in full agreement with the that of the starting material (Table 3). Sixty-six per cent of arabinose residues were (1→5)-linked indicating a rather linear arabinan. The high amount of terminally linked arabinose residues (20%) could not be explained by the number of branched arabinose residues. So, terminally linked arabinose residues were probably also present in the highly branched galactan side chains or as rather short arabinose chains. Rhamnose residues were terminally, (1→2)- or (1→2,4)-linked, while galactose were involved in almost all of the linkages possible. Xylose occurred mainly terminally linked, indicating a xylogalacturonan.

## 4. Discussion

According to the sugar composition and linkages of the (modified) hairy regions from lemon pectin, they present the structure of a rhamnogalacturonan rich in linear arabinan and branched arabinogalactan and galactan side chains. Small segments of xylogalacturonan were also found [7]. In the plant cell walls the arabinogalactan side chains of pectins may connect the rhamnogalacturonan backbone of pectins to other cell wall materials such as hemicelluloses and cellulose [18].

Yamada *et al.* [19] purified two pectins (bupleuran 2IIb and 2IIc) from the roots of *Bupleurum falcatum*. Bupleuran 2IIb has been characterised as a pectin composed of a large proportion of α-D-(1→4)-galacturonan in addition to small proportions of a rhamnogalacturonan II-like region and an endopolygalacturonase resistant region [20]. The immunostimulating activity of bupleuran 2IIb is enhanced after treatment with an endopolygalacturonase, indicating that the enzyme resistant regions, containing neutral sugar side chains such as arabinan and arabinogalactan, are important for the activity [1].

Bupleuran 2IIc, which also contained ramified regions, has anti-ulcer activity, since its oral administration prevents the formation of gastric lesions [21].

Digestion with endopolygalacturonase after de-esterification of several pectins from *Angelica acutiloba* also yields the ramified regions. The ramified region from each pectin has a more potent complement-activating activity than the corresponding original pectin [22,23,24]. Because (1→6) and (1→3,6)-linked galactosyl chains obtained from ramified regions show significant activity, the type of the neutral carbohydrate side chains attached to the rhamnogalacturonan core might be essential for expression of the complement-activating activity.

Yamada *et al.* [25] isolated a polysaccharide from the roots of *Angelica acutiloba* with a potent anti-tumour activity. This polysaccharide was characterised to contain a rhamnogalacturonan moiety, a highly branched 3,5-arabinan and a (1→4)-galactan.

The anti-complementary activity of a pectin type polysaccharide from *Plantago major* L. leaves was found to be associated to the highest molecular weight hairy region fraction, with 1,3,6 linked galactose side chains [26].

From these previously reported results on the structure of bioactive pectins from several sources it can be summarised that the distinct activities are due to rhamnogalacturonans rich in neutral sugar side chains such as arabinan, galactans and arabinogalactan region of the pectins. Reported activities seem to correlate with the presence of specific linkages such as (1→4)-, (1→6)- and (1→3,6)-linked galactosyl and branched 3,5-arabinan. The (modified) hairy regions of pectins also contain rhamnogalacturonan segments rich in neutral sugars side chains of arabinan, galactans and arabinogalactans, while also (1→6)- and (1→3,6)-linked galactosyl residues are present. Based on these findings, it could be anticipated that some immunostimulating, complement-activating and/or anti-ulcer activity may be present in these pectic fractions from lemon. Lemon hairy regions do not contain branched 3,5-arabinans which structure is mentioned in relation with anti-tumour activity. Also the methyl ester distribution in the hairy region fractions may have an effect on the bioactivity, as reported by Kiyohara and Yamada [27], who stated that the modulation between the classical and the alternative pathway of activation of the complement system may be controlled by the distribution of the methyl-ester on the ramified regions. Pectic hairy regions of lemon fruits may represent a polysaccharide with potential bioactivity. Within the near future, a detailed study towards the possible bioactivity of the various fractions would be of great interest.

**Acknowledgements**

We thank the Commission of the European Union for financial help with this research and for the financial help to J.M.R. as a fellow researcher within the "Human Capital and Mobility" programme.

**References**

[ 1] Yamada, H. (1996). In *Pectins and pectinases*, eds. Visser, J. & Voragen, A.G.J., Elsevier Science B.V., Amsterdam, p. 173.

[ 2] Yamada, H. & Kiyohara, H. (1999). In *Immunomodulatory agents from plants*, ed. Wagner, H., Birkhauser Verlag, Basel, p. 161.

[ 3] Voragen, A.G.J., Pilnik, W., Thibault, J.-F., Axelos, M.A.V. & Renard, C.M.G.C. (1995). In *Food polysaccharides and their applications*, ed. Stephen, A.M., Marcel Dekker, Inc., New York, p. 287.

[ 4] de Vries, J.A., Rombouts, F.M., Voragen, A.G.J. & Pilnik, W. (1984). *Carbohydr. Polym.*, 4, 89.

[ 5] Ralet, M.C. & Thibault, J.-F. (1994). *Carbohydr. Res.*, 260, 283.

[ 6] Ros, J.M., Schols, H.A. & Voragen, A.G.J. (1996). *Carbohydr. Res.*, 282, 271.

[ 7] Ros, J.M., Schols, H.A. & Voragen, A.G.J. (1998). *Carbohydr. Polym.*, 37, 159.

[ 8] Behall, K. & Reiser, S. (1986). In *Chemistry and function of pectins*, eds. Fishman, M.L. & Jen, J.J., ACS Symposium Series 310, Washington, p. 248.

[ 9] Hoagland, P.D. & Pfeffer, P.E. (1986). In *Chemistry and function of pectins*, eds. Fishman, M.L. & Jen, J.J., ACS Symposium Series 310, Washington, p. 266.

[10] Endress, H.U. (1991). In *The chemistry and technology of pectin*, ed. Walter, R.H., Academic Press, San Diego, p. 251.

[11] Edwards, C.A. & Parrett, A.M. (1996). In *Carbohydrates in food*, ed. Eliasson, A.C., Marcel Dekker, New York, p. 319.

[13] Schols, H.A. & Voragen, A.G.J. (1994). *Carbohydr. Res.*, 256, 83.

[12] Schols, H.A., Vierhuis, E., Bakx, E.J. & Voragen, A.G.J. (1995). *Carbohydr. Res.*, 275, 343.

[14] Ahmed, A.E. & Labavitch, J.M. (1977). *J. Food Biochem.*, 1, 361.

[15] Blakeney, A.B., Harris, P.J., Henry, R.J. & Stone, B.A. (1983). *Carbohydr. Res.*, 113, 291.

[16] Voragen, A.G.J., Schols, H.A. & Pilnik, W. (1986). *Food Hydrocoll.*, 1, 65.

[17] Schols, H.A., Posthumus, M.A. & Voragen, A.G.J. (1990). *Carbohydr. Res.*, 206, 117.

[18] Hwang, J., Pyun, Y.R. & Kokini, J.L. (1993). *Food Hydrocoll.*, 7, 39.

[19] Yamada, H., Ra, K.S., Kiyohara, H., Cyong, J.C. & Otsuka, Y. (1989). *Carbohydr. Res.*, 189, 209.

[20] Matsumoto, T., Hirano, M., Kiyohara, H. & Yamada, H. (1995). *Carbohydr. Res.*, 270, 221.

[21] Sun, X.B., Matsumoto, T. & Yamada, H. (1991). *J. Pharm. Pharmacol.*, 43, 699.

[22] Yamada, H., Kiyohara, H., Cyong, J.C. & Otsuka, Y. (1984). *Molec. Immunol.*, 22, 295.

[23] Yamada, H., Kiyohara, H., Cyong, J.C. & Otsuka, Y. (1987). *Carbohydr. Res.*, 159, 275.

[24] Kiyohara, H., Cyong, J.C. & Yamada, H. (1988). *Carbohydr. Res.*, 182, 259.

[25] Yamada, H., Komiyama, K, Kiyohara, H., Cyong, J.C., Hirakawa, Y. & Otsuka, Y. (1990). *Planta Med.*, 56, 182.

[26] Samuelsen, A.B., Smestad-Paulsen, B., Wold, J.K., Otsuka, H., Kiyohara, H., Yamada, H. & Knutsen, S.H. (1996). *Carbohydr. Polym.*, 30, 37.

[27] Kiyohara, H. & Yamada, H. (1994). *Carbohydr. Polym.*, 25, 117.

Chapter 11

# Enzymes as tools for structural studies of pectins

A.G.J. VORAGEN, P.J.H. DAAS and H.A. SCHOLS.
*Department of Food Technology and Nutritional Sciences, Food Science Group, Bomenweg 2, 6703 HD Wageningen, The Netherlands.*

Key words: pectolytic enzymes, pectin, rhamnogalacturonan, hairy regions, structure elucidation

Abstract: The structure elucidation of pectic polysaccharides has obtained new momentum and new dimensions by the increasing availability of pure, well characterised pectin modifying or depolymerizing enzymes and the discovery of a new class of enzymes active towards rhamnogalacturonan structures. They provide us with fragments which fit better in the range of modern NMR and mass spectrometric techniques. Homogalacturonan degrading enzymes give valuable information on the distribution of substituents along the galaturonan backbone.

## 1. Introduction

Pectins are important components of plant cell walls. As such they contribute to many of the functions cell walls perform in plants e.g. physiological in growth, determining cell size and shape, integrity and rigidity of tissues, ion transport, water binding, protection against infections by plant pathogens and wounding.

In edible parts of plants they determine to a large extent quality attributes of fresh fruits and vegetables (e.g. ripeness, texture), their processing characteristics in manufacture of foods (juices, nectars, purees, preserves), extractability of important constituents of plant raw materials like sugar, oil, proteins, colorants, anti-oxidants etc.

Extracted from suitable plant material (by-products like citrus peel and apple pomace) they are used in the food industry as a natural ingredient due to their ability to form gels at low concentrations and to increase the



*B.S. Paulsen (ed.), Bioactive Carbohydrate Polymers,* 129–145.

viscosity of liquid foods. They are also widely applied as stabilisers in acid milk products and recently also as fat mimetic [1].

Relevant to this symposium is their nutritional and pharmaceutical significance. They are considered as dietary fibre and have been shown to lower blood cholesterol levels. They also are claimed to have pharmaceutical activities like anti-diarrhoea, detoxicant, regulation and protection of the gastrointestinal tract, immune-stimulating activity, anti-metastasis activity, anti-ulcer activity and anti-nephrosis activity [1].

## 2. Chemical structure in general

The pectic polysaccharides are probably the most complex class of plant cell wall polysaccharides and comprise a family of acidic polymers like homogalacturonans, rhamnogalacturonans with different types of neutral polymers like arabinans, galactans and arabinogalactans attached to it, xylogalacturonans and apiogalacturonans [1,2]. Native pectins are believed to consist of a backbone in which "smooth" galacturonan regions of α-(1→4)-linked D-galacturonosyl residues are interrupted by ramified rhamnogalacturonan (hairy) regions (Figure 1) consisting of a backbone of alternating α-(1→2)-linked L-rhamnosyl and α-(1→4)-linked D-galacturonosyl residues (also indicated as rhamnogalacturonan I). Side chains are predominantly attached to 0-4 of the rhamnosyl residues. The presence of at least 30 different side chains has been indicated for rhamnogalacturonan isolated from suspension-cultured sycamore cells. Some of these side chains consisted of at least 15 glycosyl residues. Approximately one half of the (1→2) linked L-rhamnosyl residues are branched at 0-4 with side chains averaging about seven glycosyl residues in length composed of D-galactosyl and L-arabinosyl residues. Small amounts of L-fucosyl residues linked to L-rhamnosyl were also [3].

The proportion of "smooth" to "hairy" region can vary greatly depending on the type of tissue or its developmental stage.

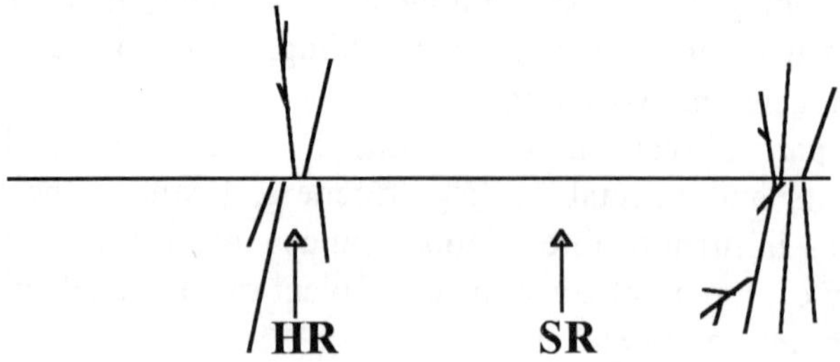


*Figure 1. Schematic structure of apple pectin including rhamnogalacturonan segments of the backbone ( smooth regions: SR) and ramified or hairy regions (HR)*

A minor component of plant cell walls is rhamnogalacturonan-II which has an extremely complex structure. It has a "homogalacturonan" backbone composed of about nine α-1,4-linked D-galacturonosyl acid residues to which four different complex side chains are attached to 0-2 or 0-3 of four of the backbone residues. Characteristic for RG-II is the presence of rare sugars like 2-0-methyl-fucose, 2-0-methyl-xylose, apiose, aceric acid, 2-keto-3-deoxy-D-manno-octulosonic acid (KDO) and 3-deoxy-D-lyxo-2-heptulosaric acid (DHA) [3,12].

By acid extraction of plant tissue particularly homogalacturonans are obtained [1]. Many of the carboxyl groups (~70%) of the galacturonosyl residues in homogalacturonans as well as in rhamnogalacturonans are methyl-esterified. The presence of acetyl groups is not a general feature of homogalacturonans, they occur in certain tissues (sugar-beet, pears). Homogalacturonans are renowned for their ability to form gels, a property widely utilised in the food industry. The gelling properties greatly depend from the degree of methyl esterification, high methoxyl pectins (>50%) forming gels in acid conditions and in the presence of sugar (>65%w/w). The higher the methoxyl content the faster the gels will form upon cooling. Low methoxyl pectins form gels with $Ca^{++}$-ions. The presence of acetyl groups prevents gel formation with $Ca^{++}$-ions [1].

Xylogalacturonans and apiogalacturonans occur in certain tissues of plants having a galacturonan backbone in common with the homogalacturonans [2,4,5,6]. Xylogalacturonans will be discussed in a later section.

To understand the many functionality's of pectin fractions it is essential to know their chemical structure which strongly depends on the origin of the pectin (Table 1). Important characteristics of pectins are galacturonic acid content, neutral sugar (linkage) composition, amount and distribution of methyl esters and acetyl groups and their molecular weight. In addition, the ratio between smooth and hairy regions, the length and chemical fine structure of the neutral side chains and the distribution of the various subunits over the backbone may influence the pectins properties as well.

*Table 1. Characteristics of pectins extracted from different sources*

| Sugar | Apple | Lemon | Sugar beet |
|---|---|---|---|
| Rhamnose | 2 | 2 | 2 |
| Arabinose | 9 | 3 | 8 |
| Xylose | 1 | tr | tr |
| Mannose | tr | tr | tr |
| Galactose | 6 | 6 | 5 |
| Glucose | 16 | 1 | 1 |
| Galacturonic acid | 66 | 88 | 84 |
| DM | 74 | 72 | 60 |
| DA | 5 | 1 | 34 |

Many of these characteristics can be determined by conventional chemical methods [1], however pectin-degrading enzymes in pure form in combination with modern spectroscopic techniques have opened new avenues in elucidating their fine structure [7,8]. In the past years a whole new class of enzymes has been found which specifically are able to degrade rhamnogalacturonan and xylogalacturonan structures [9,10] and prove to be very helpful in studying the structure of the "hairy regions". The use of enzymes also enabled us to establish the distribution of esterified and non-esterified galacturonosyl residues in commercial pectins.

## 3. Enzymes active on pectins

Two families of enzymes can be distinguished which act on pectic polysaccharides: homogalacturonan-modifying enzymes depicted in Figure 2 and rhamnogalacturonan degrading enzymes depicted in Figure 3 and will be discussed separately.

*Homogalacturonan modifying* enzymes comprise pectin methyl esterases (PME) releasing methanol from high methoxyl pectins in a blockwise (PME's from plant origin, pH optimum around 7) or random fashion (PME's from fungal origin, pH optimum around 4-5) and pectin acetyl esterases releasing acetic acid from acetylated pectins (occurring in plants and fungi). Another class are the pectin depolymerizing enzymes. Polygalacturonase (endo-PG) degrades glycosidic linkages by hydrolysis in non-esterified regions of homogalacturonans in an endo-fashion releasing mono-, di- and oligogalacturonic acids or in a exo-fashion releasing mono- or digalacturonic acid from the non-reducing end (exo-PG). Pectate lyase degrades non-esterified regions of homogalacturonans by β-elimination in an endo-fashion (endo-pectate lyase, endo-PAL) releasing 4,5 unsaturated di-, tri- and higher oligogalacturonic acids or exo-fashion releasing 4,5 unsaturated digalacturonic acid from the non-reducing end (exo-pectate lyase, exo-PAL). Pectin lyase degrades highly methyl esterified homogalacturonan regions by β-elimination in a random fashion releasing 4.5 unsaturated methyl oligogalacturonates (endo-PL, produced by fungi, pH optimum around 6). PAL's are produced by aerobic and anaerobic bacteria and have an absolute requirement for $Ca^{++}$ ions, they are optimally active at pH 8.5 [11].

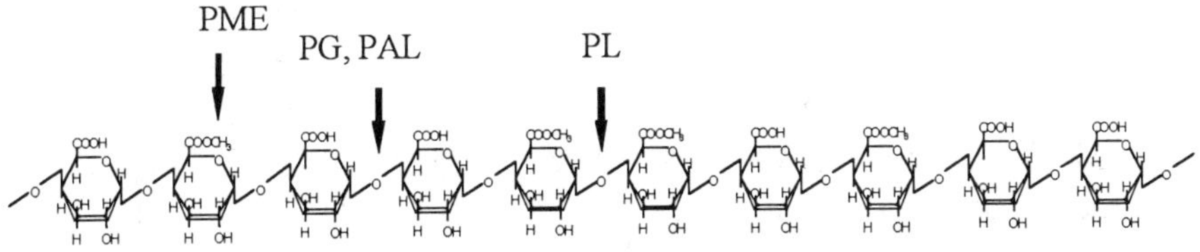


*Figure 2. Schematic structure of homogalacturonan and its enzymic modification.*

*Rhamnogalacturonan degrading enzymes* were only discovered in the last decade [13]. They comprise two different types of depolymerizing enzymes: one is a hydrolase splitting the α-GalA*p*-(1-2)-α-Rha*p* linkages in rhamnogalacturonan (RG hydrolase) while the other type is a lyase splitting the α-Rhap-(1-4)-α-GalA*p* linkage by β-elimination (RG lyase). In addition, two other enzymes were found which have a high specificity towards rhamnogalacturonan fragments: a rhamnogalacturonan rhamnohydrolase releasing rhamnosyl groups from their non-reducing end, and a rhamnogalacturonan galacturonohydrolase releasing galacturonosyl groups from the non-reducing end [14,15,16,17]. A β-galactosidase is needed for a complete degradation of rhamnogalacturonan oligomers by these exo-enzymes (Figure 3).

As an accessory enzyme for RG hydrolase and RG lyase a rhamnogalacturonan acetyl esterase (RGAE) was identified. The enzyme specifically removes acetyl groups at both 0-2 and 0-3 from galacturonosyl residues in the rhamnogalacturonan backbone and is essential for the degradation of rhamnogalacturonans by RG hydrolase or RG lyase [18]. All these enzymes were obtained from an *Aspergillus aculeatus preparation,* but they have also been found in *Aspergillus niger, Irpex lacteus, Trametes sanguinea* and *Botrytis cenera* [14].

Only recently a new type of enzyme has been isolated from *Aspergillus tubigensis* active on xylogalacturonan in an endo fashion (endo-xylogalacturonase, [10,19]). An exo-PG was also found to be active on xylogalacturonans releasing the dimer xylosyl-1,3-galacturonic acid [9,13].

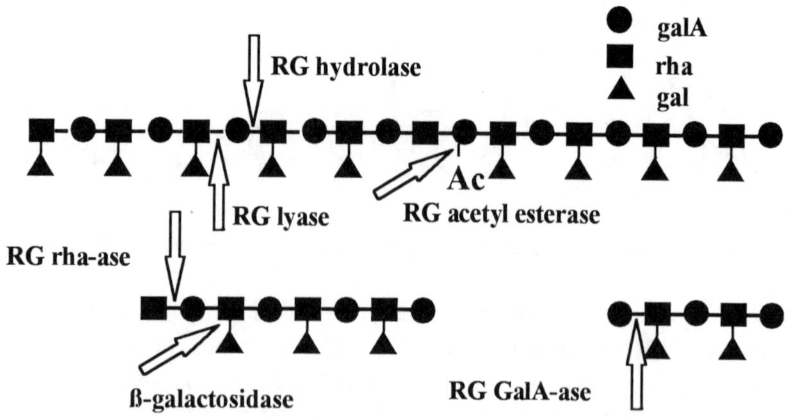


*Figure 3. Schematic representation of enzymes able to modify/degrade rhamnogalacturonans*

## 4. Strategy for structure elucidation using enzymes

The group of Albersheim was the first to use enzymes in structure elucidation of pectic material. By treatment of suspension-cultured sycamore cell walls with endo-PG and endoglucanase they solubilised RG-I and RG-II [20]. The isolation and structure elucidation of these structural units by chemical methods was subject of many publications which were conveniently reviewed by O'Neill et al. [3]. The discovery of the rhamnogalacturonan degrading enzymes gave access to rhamnogalacturonan structures and enabled their fragmentation to smaller structural units. This is described in a next section.

De Vries et al. [7]) also used enzymes in structure studies of pectins. A pre-requisite for these studies was to have homogeneous pectin fractions. This was achieved by anion-exchange chromatography and size-exclusion chromatography of the pectin preparations. By degrading these purified pectins specifically in the galacturonan backbone using enzymes active towards the (methyl-esterified) homogalacturonan region with endo-PG or endo-PAL in combination with PME or with PL and fractionation of the digests by size-exclusion chromatography, methyl oligogalacturonates were obtained next to polymeric fragments in which virtually all of the neutral sugar residues found in the parental pectin are concentrated. From these results they concluded that there is an intramolecular distribution in which the neutral sugars are concentrated in blocks of more highly substituted rhamnogalacturonan regions ("hairy"), separated by unsubstituted ("smooth") regions containing almost exclusively D-galactosyluronic residues as already illustrated in Figure 1). By degrading the purified pectins with endo-PG in the absence of PME only those parts of the homogalacturonans are degraded which are not or very low methyl-esterified. The amount and size of the oligogalacturonates obtained provide information on the distribution of methyl esters along the galacturonan backbone. This is described in more detail in one of the next sections.

## 5. Structural characterisation of hairy regions

By treatment of apple tissue with a technical enzyme preparation free of rhamnogalacturonan degrading enzymes an enzyme resistant pectic polysaccharide was obtained representing 0.26% of apple's fresh weight [21]. The dominant sugar residues present in this fraction were arabinose (55 mol%) and galacturonic acid (21 mol%), in addition to rhamnose (6%), xylose (8%), and galactose (9%). As calculated on the galacturonic acid content, the amount of methyl esters was rather moderate (DM= 42%), while the substitution with acetyl groups was rather high (DA = 60%). These characteristics resembled that of the Hairy Regions as described by De Vries et al. [22,23]. Since the fraction was obtained using an enzyme preparation containing a wide variety of activities, the fraction was named MHR (Modified Hairy Regions). Using SEC, it was shown that the MHR fraction consisted of three populations differing in molecular weight. The sugar composition of the individual populations somewhat differed: especially the high Mw fraction was richer in xylose, while the lowest Mw population contained more rhamnose and galacturonic acid residues.

Although apple MHR was resistant towards the crude enzyme mixture used, screening studies involving a large number of crude enzyme preparations revealed that an *Aspergillus aculeatus* preparation contained enzyme activities able to degrade MHR. From this preparation a new family of rhamnogalacturonan degrading enzymes were isolated and characterised by their activity towards MHR and MHR fragments (see above). Study of the degradation products formed provided valuable information on the complex structure of MHR.

Firstly, the oligomeric fragments released by RG hydrolase (only active when acetyl groups are removed by treatment with alkali or RG acetylesterase) represented alternating sequences of rhamnose and galacturonic acid residues with a single galactose residue linked to C-4 of (part of) the rhamnose moieties. This indicates that part of the rhamnogalacturonan (or RG-I) segment of apple pectin was much lower substituted with much smaller side chains as depicted before for RG-I from sycamore cells [3].

Secondly, size-exclusion chromatography of the higher Mw fragments enabled the recognition of other new subunits: an 80 kDalton xylogalacturonan segment (38 mol% xylose and 41 mol% galacturonic acid), in addition to fragments of ≈ 20-30 kDalton representing arabinans (>80 mol% of arabinose), connected to some residual stubs of the rhamnogalacturonan backbone. Anion-exchange chromatography and further characterisation of the xylogalacturonan population revealed that the single unit side chains of β-xylosyl residues were connected to O-3 of the galacturonic acid moieties. The xylose to galacturonic acid ratio of three isolated xylogalacturonan fractions varied from 0.4 to 0.7 while distinct variations exist in the methyl esterification (DM=40, 70 and 90% respectively; Schols et al., [4].

## 6. Mildly extracted pectins

Since the subunits of apple MHR have been isolated in first instance using a crude enzyme preparation containing many known and unknown enzyme activities, apple cell wall material was extracted using a mild, sequential extraction of buffers, chelating agents and diluted alkali at low temperature [5]. The findings on the structure of Hairy Regions isolated from the various pectin fractions were rather similar to those found for MHR although the relative amount of galactose was somewhat higher. RG hydrolase was able to release the typical oligomers as mentioned before, although it was observed that the amount of the oligomers released and the degree of galactose-substitution of the rhamnose moieties increased when the starting pectins were extracted under more harsh conditions (cold buffer versus diluted alkali).

When pectins from lemon albedo were extracted in a similar way, the same sub-units as mentioned for apple MHR were recognised. However, it should be stated that again the amount of the various oligomers released after RG hydrolase digestion differed. Even a fraction enriched in the xylogalacturonan subunit could be isolated (>30 mol% xyl). This was a bit surprising since the starting material before enzymic digestion and fractionation only contained ≈ 1 mol% of xylose [24,25,chapter 10 this symposiumbook].

## 7. A general model for plant cell wall pectic substances?

Using enzymes, quite some information on the structure of the ramified hairy regions of various sources [25,26,27, chapter 10 this symposium-book] has became available. The information obtained so far is depicted in a working model as shown in Figure 4.

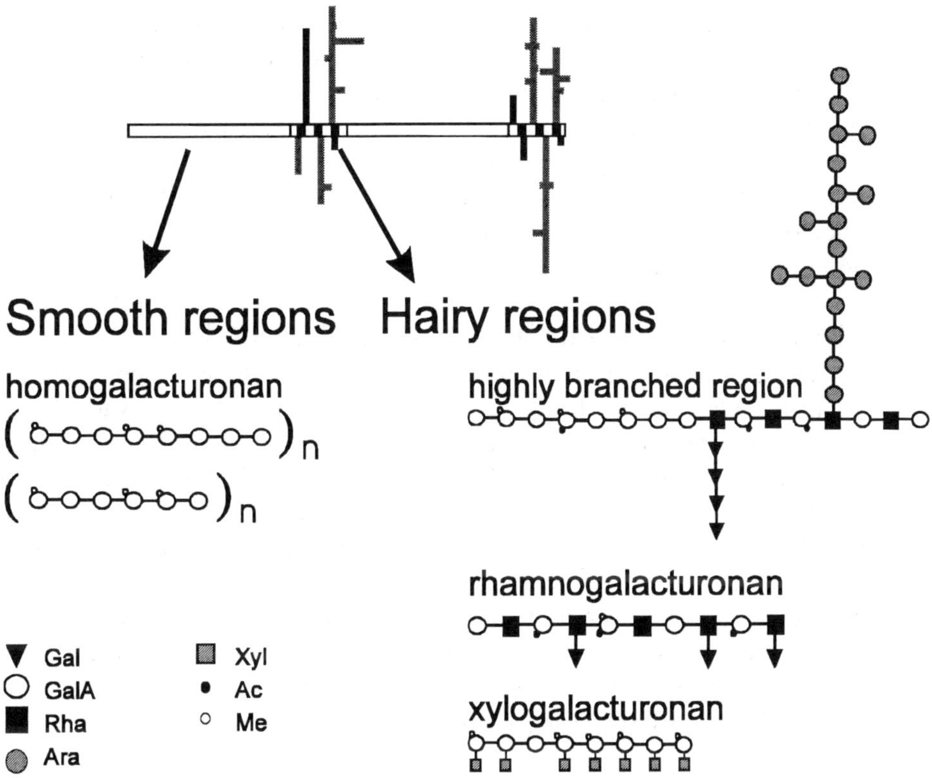


*Figure 4. Hypothetical structure of a native pectin as present in fruits and vegetables. The relative amount and arrangement of the individual subunits (including acetyl groups and methyl esters) strongly depends on the plant material of interest.*

In addition, it was questioned by ourselves whether this information could also be combined with information from literature resulting in a general framework in which most pectin structures described so far could fit. Since the first discovery of a polymeric xylogalacturonan, the presence of such a subunit has been mentioned to be present in pectins extracted from pea hulls [28],, watermelon and cotton [29] and soy beans [30].

Of course, it should be realised that, depending on the origin of the pectin, the galactose content in the neutral sugar side chains of the backbone could be much higher as given for apple pectic hairy regions at the expense of the arabinose content. Pectic subunits as described in literature such as apiogalacturonans and galactogalacturonans [2], RG-II [3,12] could not be recognised in our studies. Probably due to the origin of the cell wall material and/or the way of isolation. However, it is still believed that in general, pectins are build up from the same subunits, although it is emphasised that principal differences in relative amounts of the various subunits, their precise structure/substitution and the sequence of the subunits over a pectin molecule could be present (for possible variations within given subunits, see Schols and Voragen, [8]). An example of pectins which definitely do not contain all subunits is given by Huisman et al. [30]): no homogalacturonan regions showed to be present at all in the pectins as isolated from soybean meal. More complex rhamnogalacturonan and xylogalacturonan structures were present in higher amounts instead. Due to their high complexity, these structures are resistant to degradation by the various rhamnogalacturonanses and xylogalacturonases [30].

## 8. Bioactivity of pectins from fruit and vegetables

Not much is known on the bioactivity of pectins isolated from fruit and vegetables. On the other hand, quite some information become available for pectins isolated from several medicinal herbs which were shown to have anti-complementary activity [31]. The group of Yamada carried out extensive studies aimed at elucidating the essential structures of pectins responsible for this activity. From their work it appears that the ramified regions may have a complement-activating activity, particularly rhamnogalacturonan structures carrying β-(1→6)-linked galactose side chains, while also the RG-II like structures are being investigated thoroughly [31, chapter 2 this symposiumbook]. Also pectins from *Plantago major* L. was found to show biological activity [32].

## 9. Enzymic determination of the methyl ester distribution along the homo-galacturonan backbone

The methyl ester distribution of pectin has been the subject of study for many years. This distribution is very complex due to the fact that the methyl ester distribution should be revealed on a intramolecular level (within one molecule) and on a intermolecular level (distribution of methyl esters over various pectin molecules within a mixture [22,23]. Methods reported until 1982 have been reviewed by Taylor [33] and represent both ‘chemical’, enzymatic or combined approach. In many of these studies, rather crude enzyme preparations were used. In the early eighties, more detailed work was published by Tuerena et al. [34,35] and De Vries et al. [22,23,36]. The latter group studied the methyl ester distribution of carefully extracted apple pectin in detail using highly purified pectin lyase and endo-[22]. Comparison of the SEC patterns of the ‘native’ apple pectin digests with those of enzyme digests of a ‘transesterified’ (an assumed random esterified) apple pectin of similar DM, revealed clear dissimilarities. Fractionation of various pectin digests by HPLC anion-exchange chromatography enabled the separation and quantification of the oligomers produced according to the number of free carboxyl groups and suggested a *rather similar intramolecular* methyl ester distribution [22].). Ion-exchange fractionation of the same ‘native’ apple pectin before enzymic degradation did reveal the presence of a *non-homogeneous intermolecular* methyl ester distribution. Similar results were found for carefully extracted lemon (albedo) pectin [23]. The methyl ester distribution of commercially extracted lemon, lime, orange peel, and apple pectin was also studied using pectin lyase, and here a *non-homogeneous intermolecular* methyl ester distribution was reported [36].

Purified *endo-PG* was used by Kiyohara et al. to characterise the methyl-ester distribution of complementary activating pectic polysaccharides extracted from the root of *Angelica acutiloba* [37,38]. Separation and quantification of the saturated galacturonic acid oligomers present in the digest of pectin fractions having the same DM showed that one pectin had a rather blockwise methyl ester distribution where the other had a much more random distribution.

Following the work of De Vries, the methyl ester distribution pattern of commercially extracted lemon and apple pectins of high DM was also investigated in our group by Kravtchenko [39,40,41] using highly purified endo-PG. The intermolecular distribution of these pectins was independently determined by preparative anion-exchange and size-exclusion chromatography [39,40] prior to enzymic digestion. The latter revealed a non-homogeneous intermolecular methyl ester distribution, which mostly resembled that of a mixture of pectic polymers with blocks of non-methyl esterified galacturonic acid of various sizes. A valuable tool in the studies of Kravtchenko was HPAEC, enabling the separation and identification of uronic acid oligomers ranging from DP 1 to 17. However, because of the high pH employed during HPAEC separation, the information regarding the methyl ester content of the oligomers produced was lost [42].

Recently, we have developed two techniques that are capable of detecting the non- and partially methyl esterified galacturonic acid oligomers produced after endo-PG degradation of pectin. The first is matrix-assisted laser desorption/ionisation time-of-flight mass spectrometry (Maldi-TOF-MS; Daas et al., 1998). A Maldi-TOF-MS spectrum of an endo-PG digest of a DM30 pectin is shown in Figure 5. With this technique, finally, methyl esterified oligogalacturonides could be easily detected even when present in complex mixtures. Maldi-TOF-MS can also be used for the identification of the oligomers produced by pectin and pectate lyase [43,44].

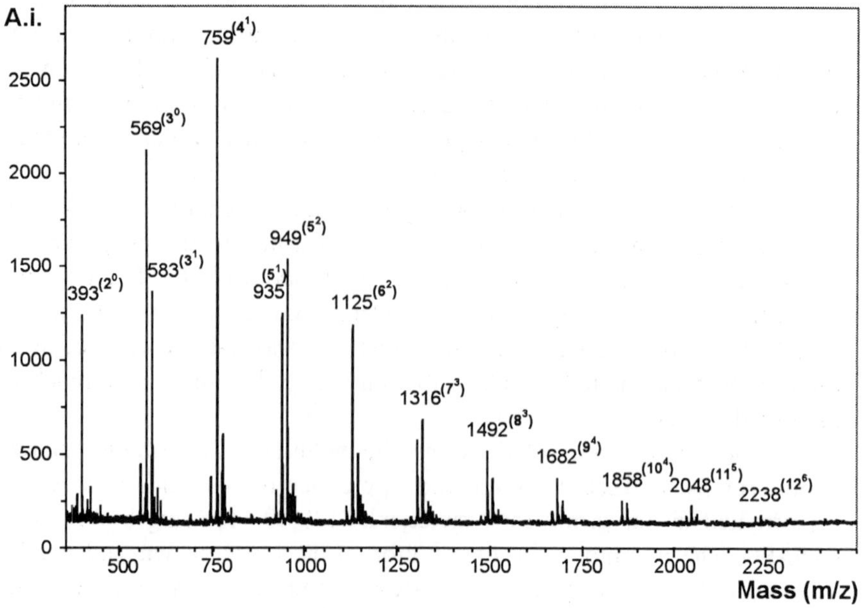


*Figure 5. Positive-ion matrix assisted laser desorption/ionisation time of flight (MALDI-TOF) mass spectrum of the endo-polygalacturonase digest of the DM 30 pectin with 2,5-dihydroxy-benzoic acid as the matrix. The masses of a selected number of peaks are shown. The numbers in parenthesis display the composition of the corresponding monosodiated galacturonic acid oligomers. The arabic number indicates the DP of the oligomer, whereas the number in superscript denotes the amount of methyl esters present. A.i., accumulated intensity.*

Unfortunately, quantification of these components is not possible with Maldi-TOF-MS and a topic for future research. The second technique we developed is HPAEC with a sodium acetate gradient at pH 5. After postcolumn sodium hydroxide addition, the non- and partially methyl esterified oligomers can be observed with a pulsed amperometric detector (PAD; [43]). The analysis of HPAEC fractions with Maldi-TOF-MS allows for an accurate identification of all components separated by the sodium acetate gradient at pH 5. After determination of the PAD-response factors, the oligomers produced can be quantified [45]. Pectins studied included a wide variety of commercial pectins as well as chemically and enzymatically de-esterified pectins with a more defined methyl ester distribution.

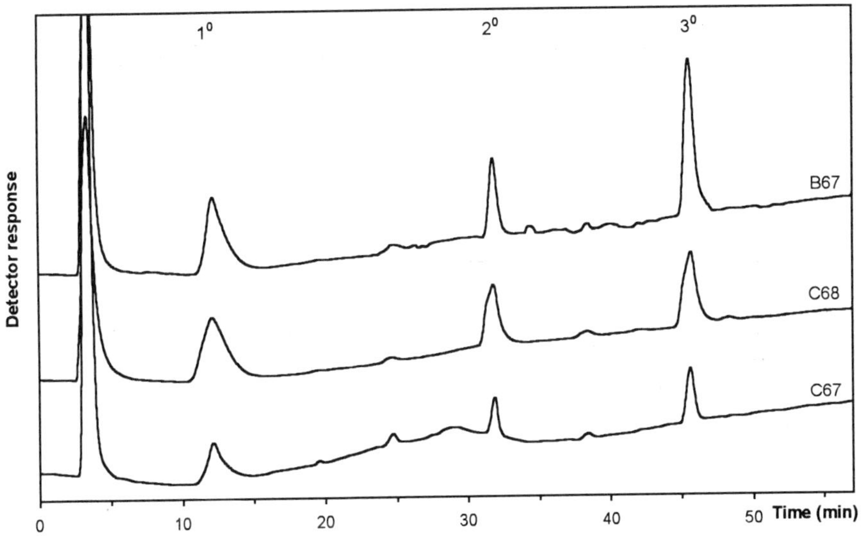


*Figure 6. HPAEC pH 5 elution patterns of three pectin endo-PG digests with a nearly identical DM (~ 67%) as indicated by the number of the sample code. Peaks corresponding to non-esterified mono-, di-, and tri-galacturonic acid are indicated.*

This -for the first time- enables the exact determination of specific structural features that could be related to the methyl ester distribution of pectin by using enzymes! Pectins that were sequentially de-esterified with tomato PME released large amounts of non-esterified galacturonic acid upon endo-PG degradation. Random methyl esterified pectins liberated the lowest amounts of non-esterified galacturonic acid, even when the DM was relatively low [45]. In general it could be concluded that the absolute amount of non-esterified mono-, di-, and tri-galacturonic acid produced after extended endo-PG degradation of pectin is an indicator for the occurrence of sequences of non-esterified galacturonic acid (so-called blocks) in pectin. The more of these blocks present and/or the larger the average size of these blocks, the more non-esterified mono-, di-, and tri-galacturonic acid produced. As an example, in Figure 6 three endo-PG digests are shown of pectins with a nearly identical DM (~67) that contained different amounts of non-esterified galacturonic acid; indicating a clearly different methyl ester distribution. For the total amount of non-esterified galacturonic acid liberated, expressed as the percentage of the total number of non-esterified galacturonic acids present in pectin, the term "degree of blockiness" (DB) was introduced [45]. In Figure 7, the different DB's that can be observed for pectins of various DM are shown. In this way pectins having the same degree of methyl esterification and different functional behaviour can be discerned. Current research is directed towards a more detailed study of all the components present in the HPAEC elution profiles and the coupling of these data with specific methyl- and/or non-methyl ester distributions in pectin.

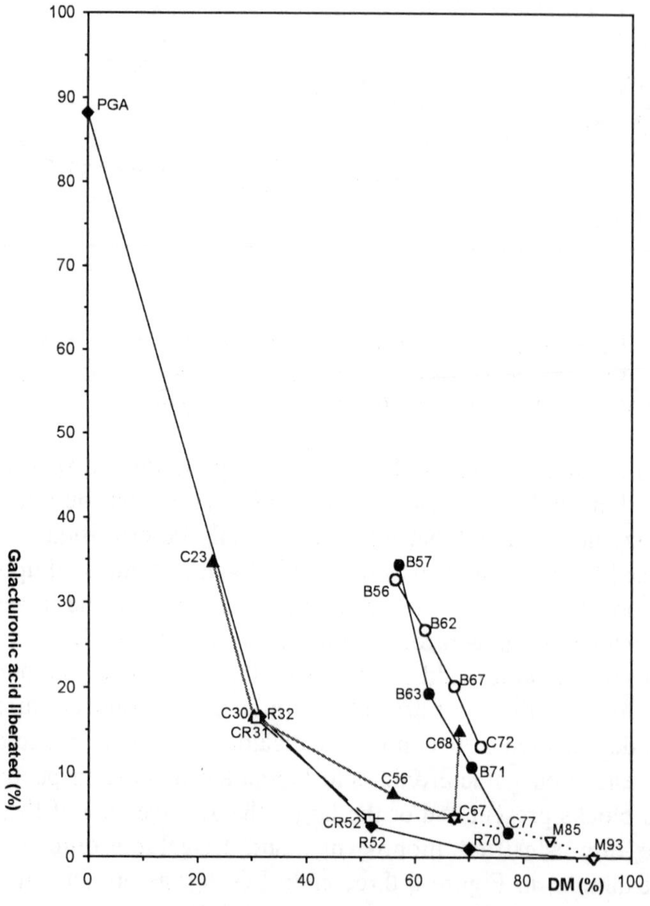


*Figure 7. Percentage of non-esterified GalA residues liberated by endo-PG versus the DM of pectin. The following pectins are shown in series: random methyl-esterified (M93,R70-R32, PGA; _◆_), alkaline de-esterified (C67,CR52,CR32; --◊--), commercial (C68-C23; _▲_), additionally methyl-esterified (C67,M85,M93; ··∇··), and tomato PME de-esterified C77 (C77,B71,B63,B57; _●_) and C72 (C72,B67,B62,B56; _O_)..*

## 10. Concluding remarks

The use of enzymes as analytical tools for structure elucidation of complex carbohydrate structures is by far not exhausted, the more so since still new enzymes active on such structures are being found. For the identification and structure elucidation of epitopes in biologically active polysaccharides, enzymes in particular can be very helpful due to their specificity.

## References

[ 1] Voragen, A.G.J., Pilnik, W., Thibault, J.-F., Axelos, M.A.V. & Renard C.M.G.C. (1995) In *Food polysaccharides and their applications*, ed. Stephen, A.M., Marcel Dekker, New York, pp. 287-339

[ 2] De Vries, J.A. (1988) In *Gums and Stabilizers for the Food Industry 4*, eds. Philips G.O., Wedlock, D.J. & Williams, P.A., IRL Press, Oxford, pp. 25-29

[ 3] O'Neill, M., Albersheim, P., & Darvill, A.G. (1990) In *Methods in Plant Biochemistry, Carbohydrates* vol. 2, ed. Dey P.M., Academic, London, pp. 415-441

[ 4] Schols, H.A., Bakx, E.J., Schipper, D. & Voragen A.G.J. (1995a) *Carbohydr. Res.*, 279, 265-279

[ 5] Schols, H.A. Vierhuis, E. Bakx, E.J. and Voragen. A.G.J. (1995b) *Carbohydr. Res.*, 275, 343-360

[6] Albersheim, P., Darvill, A.G., O'Neill, M.A., Schols, H.A. & Voragen, A.G.J. (1996) In *Progress in Biotechnology* 14, *Pectins and Pectinases*, eds. Visser, J. & Voragen A.G.J., Elsevier, Amsterdam, pp. 47-55

[ 7] De Vries, J.A., Rombouts, F.M., Voragen, A.G.J. & Pilnik, W. (1982) *Carbohydr. Polym.*, 2, 25-33

[ 8] Schols, H.A., & Voragen, A.G.J. (1996) In *Progress in Biotechnology* 14, *Pectins and Pectinases*, eds. Visser, J. & Voragen A.G.J., Elsevier, Amsterdam, pp. 3-19

[ 9] Beldman, G., Mutter, M., Searle-van Leeuwen, M.J.F., van den Broek, L.A.M., Schols, H.A. & Voragen, A.G.J. (1996a) In *Progress in Biotechnology* 14, *Pectins and Pectinases*, eds. Visser, J. & Voragen A.G.J., Elsevier, Amsterdam, pp. 231-245

[10] Meeuwsen, P.J.A. et al. (1999) Novel endo-xylogalacturonase, Patent WO-2895-N73992A

[11] Pilnik, W. & Voragen, A.G.J. (1991) In *Food Enzymology* vol. 1, ed. Fox, P.F., Elsevier, London, pp. 303-336

[12] Doco, T., Williams, P., Vidal, S. & Pellerin, P. (1997) *Carbohydr. Res.* 297 181-186

[13] Beldman, G. et al (1996b) *Biotechnol. Lett.* 18, 707-712

[14] Mutter, M. (1997) New rhamnogalacturonan degrading enzymes from *Aspergillus aculeatus*. PhD Thesis, Wageningen Agricultural University

[15] Mutter, M., Beldman, G., Schols, H.A. & Voragen, A.G.J. (1994) *Plant Physiol.*, 106, 241-250

[16] Mutter, M., Colquhoun, I.J., Schols, H.A., Beldman, G. & Voragen, A.G.J. (1996a) *Plant Physiol.* 110, 73-77

[17] Mutter, M., Renard, C.M.G.C., Beldman, G., Schols, H.A. & Voragen, A.G.J.(1996b) In *Progress in Biotechnology* 14, *Pectins and Pectinases*, eds. Visser, J. & Voragen A.G.J., Elsevier, Amsterdam, pp. 263-274.

[18] Searle-van Leeuwen, M.J.F., van den Broek, L.A.M., Schols, H.A., Beldman, G. & Voragen A.G.J. (1992) *Appl. Microbiol. Biotechnol.*, 38, 347-349

[19] Van der Vlugt-Bergmans, C.J.B., Meeuwsen, P.J.A., Voragen, A.G.J. & van Ooyen, A.J.J. (1999), *Biochim. Biophys. Acta*, submitted

[20] Keegstra, K., Talmadge, K.W., Bauer, W.D. & Albersheim, P. (1973) *Plant Physiol.* 51 188-197

[21] Schols, H.A., Posthumus, M.A. & Voragen, A.G.J. (1990) *Carbohydr. Res.*, 206, 117-129

[22] De Vries, J.A., Rombouts, F.M., Voragen, A.G.J. & Pilnik, W. (1983a) *Carbohydr. Polym.* 3, 245-258

[23] De Vries, J.A., Rombouts, F.M., Voragen, A.G.J. & Pilnik, W. (1983b) *Carbohydr. Polym.* 4, 89-10

[24] Ros, J.M., Schols, H.A. & Voragen. A.G.J. (1996) *Carbohydr. Res.* 282, 271-284

[25] Ros, J.M., Schols, H.A. & Voragen. A.G.J. (1998) *Carbohydr.* Polymers, 37, 159-166

[26] Schols, H.A. & Voragen, A.G.J. (1994) *Carbohydr. Res.*, 256, 83-95

[27] Schols, H.A., Voragen, A.G.J. & Colquhoun. I.J. (1994) *Carbohydr. Res.*, 256, 97-111

[28] Weightman, R.M. Renard, C.M.G.C. & Thibault, J.-F. (1994) *Carbohydr. Polym.* 24, 139-148

[29] Yu, L. & Mort, A.J. (1996) In *Progress in Biotechnology* 14, *Pectins and Pectinases*, eds. Visser, J. & Voragen A.G.J., Elsevier, Amsterdam, pp. 79-88

[30] Huisman, M.M.H.,Schols,H.A. & Voragen, A.G.J.(1999) *Carbohydr Polym.*38,299-307

[31] Yamada, H. & Kiyohara H. (1999) In *Immunomodulatory Agents from Plants*, ed. Wagner, H., Birhhäuser Verlag Basel, pp 161-202

[32] Samuelsen, A.B., Paulsen, B.S., Wold, J.K., Otsuko, H., Kiyohara, H., Yamada, H. & Knutsen, S.H. (1996) *Carbohydr. Polym.* 30, 37-44

[33) Taylor, A.J. (1982) *Carbohydr. Polym.* 2, 9-17

[34] Tuerena, C.E., Taylor, A.J. & Mitchell, J.R. (1981) *Carbohydr. Polym.* 2, 193-203

[35] Tuerena, C.E., Taylor, A.J. & Mitchell, J.R. (1984) *J. Sci. Food Agric.*, 35, 797-804

[36] De Vries, J.A., Hansen, M., Søderberg, J., Glahn, J. P.E. & Pedersen J.K. (1986) *Carbohydr. Polym.*, 6, 165-176

[37] Kiyohara, H., Cyong, J.-C. & Yamada, H. (1988) *Carbohydr. Res.* 182, 259-275

[38] Kiyohara, H. & Yamada, H. (1994) *Carbohydr. Polym.* 25, 117-122

[39] Kravtchenko, T.P., Berth, G., Voragen, A.G.J. & Pilnik, W. (1992a) *Carbohydr. Polym.* 18, 253-263

[40] Kravtchenko, T.P., Voragen, A.G.J. & Pilnik, W. (1992b) *Carbohydr. Polym.*, 18, 17-25

[41] Kravtchenko,T.P., Voragen, A.G.J. & Pilnik, W. (1992c) *Carbohydr. Polym.*19,115-124

[42] Kravtchenko, T.P., Penci, M.,Voragen, A.G.J. & Pilnik, W. (1993) *Carbohydr. Polym.* 20, 195-205

[43] Daas, P.J.H., Arisz, P.W., Schols, H.A., De Ruiter, G.A. & Voragen, A.G.J. (1998) *Anal. Biochem.* 257, 195-202

[44] Körner, R., Limberg, G., Dalgaard Mikkelsen, J. & Roepstorff, P. (1998) *J. Mass. Spectrom.* **33**, 836-842

[45] Daas, P.J.H., Meyer-Hansen, K., Schols, H.A., De Ruiter, G.A. & Voragen, A.G.J. (1999) *Carbohydr. Res.*, in press

# Index

# Proceedings of the Phytochemical Society of Europe

42. N.J. Kruger, S.A. Hill and R.G. Ratcliffe (eds.): *Regulation of Primary Metabolic Pathways in Plants*. 1999 ISBN 0-7923-5494-X
43. L. Bohlin and J.G. Bruhn (eds.): *Bioassay Methods in Natural Product Research and Drug Development*. 1999 ISBN 0-7923-5480-X
44. B.S. Paulsen (ed.): *Bioactive Carbohydrate Polymers*. 2000 ISBN 0-7923-6119-9

For further information about the series and how to order please visit our Website http://www.wkap.nl/series.htm/PPSE

KLUWER ACADEMIC PUBLISHERS – DORDRECHT / BOSTON / LONDON

# Proceedings of the Phytochemical Society of Europe

42. [illegible] (eds.): [illegible] 1998 ISBN 0-7923-[illegible]-X
43. [illegible] (eds.): [illegible] 1998 ISBN 0-7923-[illegible]-X
44. B.S. Paulsen (ed.): [illegible] 2000 ISBN 0-7923-6119-0

For further information about the series and how to order, please visit our Website http://www.wkap.nl/series.htm/PSE

KLUWER ACADEMIC PUBLISHERS – DORDRECHT / BOSTON / LONDON